BEST TALL BUILDINGS 2008:
CTBUH INTERNATIONAL AWARD WINNING PROJECTS

ANTONY WOOD

Council on Tall Buildings and Urban Habitat
Illinois Institute of Technology, S.R. Crown Hall
3360 South State Street
Chicago, IL, 60616
Phone: +1 (312) 909 0253
Fax: +1 (610) 419 0014
Email: info@ctbuh.org
www.ctbuh.org

College of Architecture
ILLINOIS INSTITUTE OF TECHNOLOGY

Elsevier Inc./Architectural Press
30 Corporate Drive
Suite 400
Burlington, MA 01803
USA

Linacre House
Jordan Hill
Oxford OX2 8DP
UK

Title: ***Best Tall Buildings 2008: CTBUH International Award Winning Projects***

ISBN: 978-1-85617-674-3

Editor: Antony Wood
Design and Layout: Steven Henry & Christopher Watkins
Published by: Council on Tall Buildings and Urban Habitat in conjunction with the Illinois Institute of Technology and Elsevier Inc./Architectural Press.

Printed in Canada by Transcontinental.

CTBUH has endeavored to determine the copyright holders of all images. Those uncredited have been sourced from listed authors or from within CTBUH.

On the cover: Clockwise from upper left: 51 Lime Street (Winner: Europe); The New York Times Building (Winner: Americas); Shanghai World Financial Center (Winner: Asia & Australasia); Bahrain World Trade Center (Winner: Middle East & Africa). See page 128 for image credits.

CTBUH AWARDS 2008

CONTENTS

Foreword 7

CTBUH Chair Message 8

CTBUH Awards Chair Message 9

Awards Jury 10

Acknowledgments 11

Regional Tall Building Awards

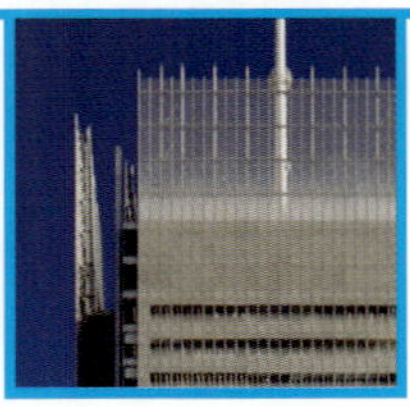

AMERICAS

Winner: The New York Times Building 16

340 on the Park 22

7 World Trade Center 24

Comcast Center 26

One Rincon Hill 28

San Francisco Federal Building 30

ASIA & AUSTRALASIA

Winner: Shanghai World Financial Center 34

181 William Street 40

AIG Tower 42

Fujian Provincial Electric and Power Company Headquarters 44

Lumière Residences 46

Newton Suites 48

Nina Tower Development 50

One Island East 52

One Raffles Quay 54

Park Tower at Beijing Yintai Centre 56

Poly International Plaza 58

The Sail @ Marina Bay 60

EUROPE

Winner: 51 Lime Street **64**
Covent Garden 70
Het Strijkijzer 72
Kanyon Complex 74
Madou Plaza 76
SyV Tower 78
Torre Espacio 80
Unity 82

MIDDLE EAST & AFRICA

Winner: Bahrain World Trade Center **86**
Aspire Tower 92
Pearl Breeze 94

Lifetime Achievement Awards

Lynn S. Beedle Award: Cesar Pelli 98
Fazlur Rahman Khan Medal: William F. Baker 104
CTBUH Fellow: Sabah Al-Rayes 110
CTBUH Fellow: James Forbes 111

Council on Tall Buildings and Urban Habitat

Review of the 2007 CTBUH Awards & Awards Dinner 112
CTBUH Height Criteria 116
100 Tallest Buildings in the World 117
CTBUH Organizational Structure 123
CTBUH Organizational Members 124

Index by Buildings 125
Index by Companies 126
Image Credits 128

FOREWORD

It is with great pleasure that we publish this inaugural edition of the CTBUH Annual International Award Winning Projects publication, profiling the very best in the realization of tall buildings from the past 12 months or so. Our Awards have grown in both number and stature since their launch in 2002. Last year we added two new "Best Tall Building" awards to recognize the team achievement in delivering any tall building of quality, and to compliment the established lifetime achievement Lynn S. Beedle and Fazlur Rahman Khan awards.

This year we extend the awards again to recognize design and construction excellence in all geographic regions of the world. One of the four regional tall building winners highlighted in these pages will also be awarded the overall title of "Best Tall Building 2008", to be announced at the CTBUH annual awards dinner in November 2008.

More than recognizing the achievements of a small number of very special projects though, what this book achieves is delivering an overview of the current directions and trends in tall building design. Whilst the commercially efficient and sculpturally iconic design approaches for tall buildings are still clearly evident across all regions of the globe, the trend towards more environmental towers—and a resultant new aesthetic—is now emerging.

Thus from the dominant wind turbines of the Bahrain World Trade Center to the sublime horizontal ceramic façade rods of The New York Times Building; from the material-saving structural design solution of the Shanghai World Financial Center to the deep envelope fins of 51 Lime Street, we now see sustainability affecting the form and expression of the building, and a new architectural language emerging. This is true not only in the winners profiled here, but also in several of the nominees—witness the spectacular 'hanging gardens' of Singapore's Newton Suites, or the green roofs and rainwater-recycling systems of Chicago's 340 on the Park.

These buildings are the products of those clients and consultant teams who recognized the need to start addressing the challenges of climate change four or five years ago, even before "green" entered the mainstream vocabulary. Though history may deem some of these approaches as being ultimately sustainable "dead ends", all should be applauded for their bravery and vision to move far beyond the commercial norm. The industry needs to better recognize that we are still in an experimental stage in determining what works and what doesn't work in green building design generally, and better support the projects which are at least exploring the possibilities. This book, and the awards that inform them, are certainly intended to do that. These emergent sustainable-technological and environment-aesthetic trends will be increasingly apparent in these pages in the years to come.

Antony Wood
Editor
CTBUH Executive Director
Chicago, Sept. 2008

CTBUH CHAIR MESSAGE

This is the first year that the Council has expanded its awards to include awards for tall buildings in all regions of the world. The increase in awards is a reflection of the ongoing growth of the Council and the increasing interest in tall buildings from around the world. We now have tall building awards for Americas, Asia & Australasia, Europe and the Middle East & Africa. Within each region there is an amazing selection of quality tall buildings.

We have been extremely pleased with the response to our invitation for awards submission; so much so that we decided to publish the results in this "Best Tall Buildings 2008: CTBUH Awards Book", which will appear on bookshelves around the world. The world has been on a tall building construction boom in recent years and there is a very exciting crop of tall buildings recently completed, and under construction, that will be changing skylines around the world.

Last year the Council made awards for the Best Tall Building and the Best Sustainable Building, and as you can imagine this created an enormous debate about how could the best building not also be the most sustainable building. In our resource-limited world sustainability is certainly one of the main drivers of tall building design. The Council expects that the level of sustainability in the best projects will continue to increase, in line with the level of sustainability being achieved in the industry.

Our 8th World Congress held in Dubai in March of this year, titled "Tall and Green", was our first effort to try to define sustainability in tall buildings. This conference was one of our most successful ever and that was a start. I am very excited by the progress of our Sustainability Working Group, and with great support from our members and other sister organizations we are now making big strides in this area.

Finally I would like to congratulate all the winners and all the runners up. The design and construction of these tall buildings is an enormous collaborative effort between architects, engineers, builders and developers; and most of us involved in the industry are extremely proud, pleased and committed to giving our best to these projects. I would particularly like to congratulate Cesar Pelli and Bill Baker for their overall and individual achievements in the tall building industry. Both Cesar and Bill have helped to shape the skylines of the world and we are pleased to recognize this with the Lynn S. Beedle Award and the Fazlur Rahman Khan Medal, respectively.

I am also particularly pleased to congratulate Sabah Al-Rayes and Jim Forbes who have been elected as Fellows of the Council in recognition of the hard work that they have given to the Council and the many years of support.

Congratulations to all those who have participated in these projects!

David Scott
CTBUH Chairman
Arup
New York City, Sept. 2008

CTBUH AWARDS CHAIR MESSAGE

The CTBUH Awards program is our way to recognize extraordinary individuals, through our two lifetime achievement awards, and buildings, through the best tall building awards. These award winners have added significantly to the design and realization of tall buildings and have profoundly shaped the urban environment. The awards program carries out the mission of the Council by bringing to the surface significant aspects of the tall building industry. Recognizing the hard work of individuals and teams is most evident; however, this recognition also makes information and data available to help others gain knowledge and understanding of tall buildings and the urban habitat, therefore elevating the high standards of the profession.

We strive to canvas the globe to look far and wide for significant contributions to tall building design and their contribution to the vitality of urban environments. This is evident in this year's expansion of the "Best Tall Building" award to identify specific regions. This focus had tremendous results! We had extraordinary interest in the awards program this year and received more, high quality, submissions than any other year. We appreciate the growing participation in the CTBUH Awards Program. It is certainly a sign of the rising importance of the Council and the dramatic growth of tall building design around the world.

Each year the Council selects a diverse spectrum of design and construction industry leaders to review and jury the submissions. Architects, engineers, sustainable design experts, and academicians from all regions of the world were part of this year's jury. Now more than any time in history, high rise structures play a significant role in the identity and vitality of cities and have the ability to protect our precious natural environment. The jury discussed major themes to evaluate the nominees for the individual awards as well as the projects. Education, collaboration, integration, sustainability, engaging the urban habitat, and the ability to transform the client and program beyond economic drivers to create powerful content-rich design to serve society were outlined as points to evaluate the submissions.

At the conclusion of an intense and rigorous jury process we believe this year's award winners represent an outstanding contribution to the profession and society. We hope that this body of work will create further interest in tall buildings and inspire the next generation of extraordinary structures and cities around the world.

To conclude I would like to personally thank my esteemed award committee colleagues who volunteered their time and intelligence to assist in debating the submissions and selecting our winners.

CONGRATULATIONS to the 2008 award winners!

Tim Johnson
Awards Committee Chairman
NBBJ
New York City, Sept. 2008

AWARDS JURY

Tim Johnson | *NBBJ* | New York City, USA

Tim Johnson is a partner with NBBJ, serving as design lead for the New York office, while providing strategic direction for the firm's international corporate and commercial practice. Through his award-winning work—ranging from corporate towers and hospitals, to interiors, educational institutions, and transportation facilities—Johnson leverages social and cultural environments to redefine the urban experience and to improve the quality of life. Prior to joining NBBJ in 1996, Johnson received a Bachelor of Architecture from the University of Minnesota, and a Master of Science in Advanced Architecture Design from Columbia University in New York.

Fiona Cousins | *Arup* | New York City, USA

Fiona is a Principal at Arup and currently chairs the US Green Building Council's New York chapter. She has extensive experience in project management, mechanical engineering and sustainability for both corporate and institutional clients with a background as a mechanical engineer. Fiona received both her MSt in Interdisciplinary Design for the Built Environment and a BA in Engineering Science (Honors) from the University of Cambridge. She has been a LEED Accredited Professional since 2001.

James Forbes | *Hyder Consulting Pty. Ltd.* | Sydney, Australia

Jim Forbes has been involved with the planning, design, co-ordination and supervision of a wide variety of buildings and engineering structures for over forty years. His acknowledged expertise in concrete and concrete structures led him towards a strong involvement with the Concrete Institute of Australia in which he has held many positions including Federal President. Jim was invited to join the Presidium of the Federation Internationale du Beton in 1998 and was elected President of this prestigious international body from 2002 to 2004.

Dan Sesil | *Leslie E. Robertson Associates* | New York City, USA

Daniel A. Sesil, P.E., S.E., is a Partner at Leslie E. Robertson Associates (LERA). With the firm since 1983, Mr. Sesil has worked on several of LERA's most widely recognized civic and cultural projects. He has extensive experience in the design of special atrium structures and long span, column free spaces required for museums, galleries, and visitor centers. Mr. Sesil has served as Adjunct Assistant Professor at Columbia University. He holds a Master of Civil Engineering from Purdue University, and earned a Bachelor of Civil Engineering from Marquette University.

Carol Willis | *The Skyscraper Museum* | New York City, USA

Carol Willis is the founder, director, and curator of The Skyscraper Museum. An architectural and urban historian, she has researched, taught, and written about the history of American city building. Ms. Willis has been an Adjunct Associate Professor of Urban Studies at Columbia University since 1989. As an undergraduate, Ms. Willis majored in Art History at Boston University, graduating magna cum laude and Phi Beta Kappa in 1971. She did her graduate work in architectural history at Columbia University in the Department of Art History and Archaeology, receiving an M.A. in 1976 and M.Phil. in 1979.

Acknowledgments:

We would like to extend our gratitude to our awards committee chairman and awards jurors who volunteered their time and expertise to deliberate on our nominees and choose our winners.

We would also like to thank all our winners and nominees whose generous contributions in time and material made this inaugural book possible.

BEST TALL BUILDING

AWARDS 2008

AWARD CRITERIA

These awards recognize projects that have made extraordinary contributions to the advancement of tall buildings and the urban environment, and that achieve sustainability at the highest and broadest level. The projects must also exhibit processes and/or innovations that have added to the profession of design and enhance the cities and the lives of their inhabitants. Criteria for submission include:

1. The project must be completed and occupied within the past two years.

2. The project advances seamless integration of architectural form, structure, building systems, sustainable design strategies, and life safety for its occupants.

3. The project must exhibit sustainable qualities at a broad level:

   Environmental- Minimize effects on the natural environment through proper site utilization, innovative uses of materials, energy reduction, use of alternative energy sources, reduced emissions and water consumption.
   People- Must have a positive effect on the inhabitants and the quality of human life.
   Community- Must demonstrate relevance to the contemporary and future needs of the community in which it is located.
   Economy- The building should add economic vitality to its occupants, owner and community.

4. The project must achieve a high standard of excellence and quality in its realization.

5. The site planning and response to its immediate context must ensure rich and meaningful urban environments.

6. The contributions of the project should be generally consistent with the values and mission of the CTBUH.

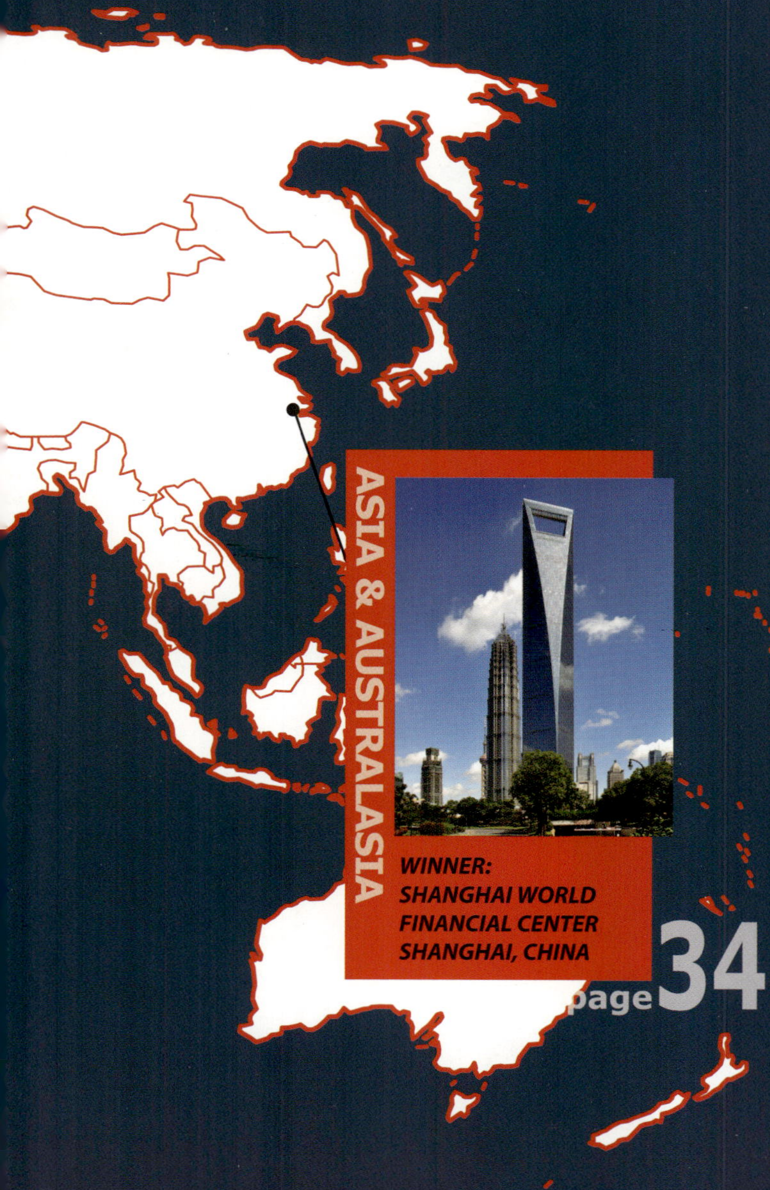

BEST TALL BUILDING

NOMINEE:
340 ON THE PARK
CHICAGO
USA

page 22

NOMINEE:
ONE RINCON HILL
SAN FRANCISCO
USA

page 28

NOMINEE:
SAN FRANCISCO
FEDERAL BUILDING
SAN FRANCISCO, USA

page 30

AWARDS 2008

AMERICAS

WINNER

WINNER:
THE NEW YORK TIMES BUILDING
NEW YORK CITY, USA

page 16

NOMINEE:
7 WORLD TRADE CENTER
NEW YORK CITY, USA

page 24

NOMINEE:
COMCAST CENTER
PHILADELPHIA
USA

page 26

2008

WINNER:

BEST TALL BUILDING AMERICAS

THE NEW YORK TIMES BUILDING

NEW YORK CITY, USA

AWARDS JURY / EDITOR STATEMENT

The New York Times Building is an important new addition to the New York skyline, but, for the Times Company, the building needed to be more than just a beautiful building. It had to support the dramatic transformation of this venerable institution as it reinvented itself in the face of profound shifts in media and market. Indeed, as the publisher repeatedly pushed, the building needed to change the way the Company worked, and this goal suffused the development of the design.

The New York Times Building incorporates many transcendental themes in good architecture—volume, views, light, respect for context, relationship to the street—with a design that is open and inviting, providing its occupants with a sense of the city around them. The resulting building treads lightly on the natural environment and is an affirmation of the Times Company's commitment to the city, its Times Square neighborhood, and to the transformative power of great architecture.

The Company's interior design creates the highest quality interior environment for a 21st century media company, ensuring productivity and that long-term operational and workplace health needs are not only met, but are exceeded. A challenge of the skyscraper is reducing heat from the sun, and the two typical methods are smaller windows or heavily coated glass, methods that, in the words of the building's architect, produce "selfish buildings," where the views and light are compromised for both pedestrians looking into the building and occupants looking out.

In contrast to the opaque design of many urban office buildings, The New York Times Building achieves a high level of transparency with the innovation of a second skin of cleverly spaced ceramic rods to reduce the heat load to a point where the building is energy efficient and yet has the great luxury of floor-to-ceiling, water-white glass.

The result from the outside is a unique level of transparency to the street—revealing the activity within—which embodies the Company's mission of transmitting an unclouded, lucid report of the news to its public (*see image on page 18*). The result from the inside is a strong connection with the City and a remarkable degree of natural light. But such a wealth of light also

"In the waning days of hermetically sealed, form-driven towers sheathed in glass, The New York Times Building is a refreshing example of a thoughtful, sustainable, and beautiful box."

-Tim Johnson, NBBJ

The west façade of The New York Times Building; image © David Sundberg/Esto

PROJECT DETAILS

Completion Date
July 1, 2007

Height
319 meters/1,046 feet

Total Area
463,601 square meters/
1,521,000 square feet

Use
Office, Retail

Owner
The New York Times Company
Forest City Ratner Companies

Developer
Forest City Ratner Companies

Architect
Renzo Piano Building Workshop/ FXFOWLE Architects

Structural Engineer
Thornton Tomasetti

MEP Engineer
WSP Flack + Kurtz

Contractor
AMEC Construction Management (core and shell)
Turner Construction (interior)

Other Consultants
Gensler

THE NEW YORK TIMES BUILDING

NEW YORK CITY, USA

The disappearing ceramic skin seen on the east façade of the tower at night; image © David Sundberg/Esto

brings glare. To address this issue a first-of-a-kind system was developed: shades automatically adjust to block glare, and the lights dynamically adjust to dim or turn off if the natural light is sufficiently bright. Moreover, with fully digital ballasts, each group can set the level of lighting that they want in their space.

The combination of these elements produced the near-magical confluence of an improved workplace with substantial energy savings. And as an added dividend, the Times Company has widely shared the results of the research and development with the larger community, and the manufacturers have made the system part of their standard offerings.

This dedication to transparency and animated spaces is evident in the building's lobby. Resisting the typical cold, forbidding corporate lobby, the architects opened the lobby by splitting the core to create an open and inviting space with a vista through the building of some 114 meters (375 feet). This unusual approach not only lightens the building so that it gently touches the ground, but it creates the exciting experience of 'layers of transparency' from

The New York Times Company cafeteria; image © Nic Lehoux

The lobby view from west to east featuring the garden; image © Nic Lehoux

The south façade from 8th Avenue; image © David Sundberg/Esto

THE NEW YORK TIMES BUILDING

NEW YORK CITY, USA

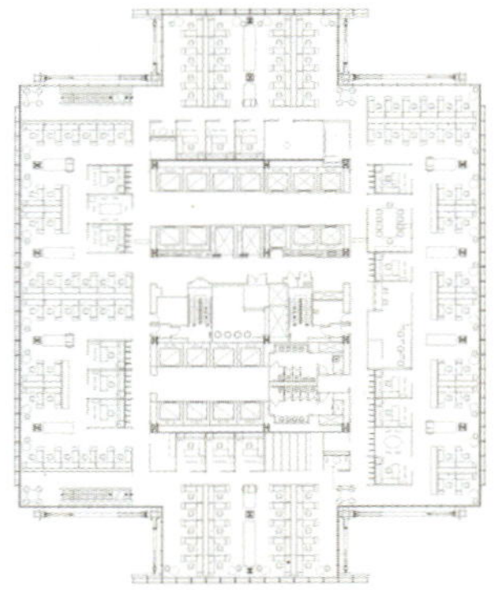

1. Retail
2. Lobby
3. Courtyard
4. The Times Center Auditorium
5. Auditorium Lobby
6. Loading Dock
7. Newsroom
8. Editorial Departments

Typical Tower Floor 18-22

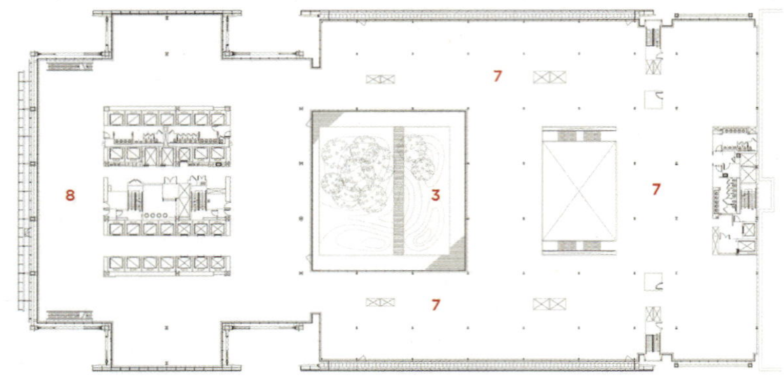

Fourth Floor

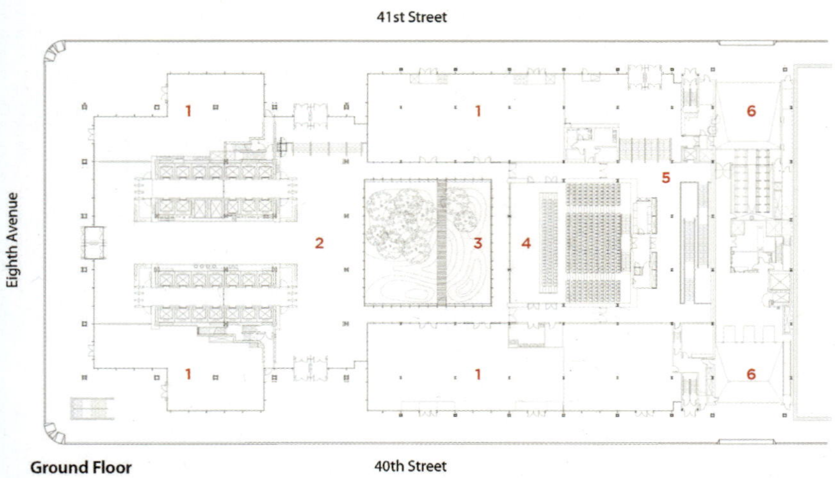

Ground Floor

Floor plans

the hustle and bustle of the lobby through to the quiet of a rare, urban 21x21x21 meter (70x70x70 foot) open-air garden featuring seven birch trees (*see ground floor plan at left and lobby view image on page 19*).

Awards Jury Statement:

In the waning days of hermetically sealed, form-driven towers sheathed in glass, The New York Times Building is a refreshing example of a thoughtful, sustainable, and beautiful box. The attention to detail at all levels is notable, from the openness that encourages the city to flow into the lobby, to the highly articulated skin which shades the building's inhabitants while still allowing an abundance of natural light, to the dematerialization of the building's top disappearing into the sky via projecting glass plates and an enormous javelin mast. The building works at all levels—the individual, the community, and on the city skyline. The expression of exposed structure at its corners gives you the impression that the building is alive and breathing. The jury was impressed by the design team's ability to balance the needs of the end-user with that of the developer. This is an extraordinary building—perhaps the Seagram Building of the 21st Century.

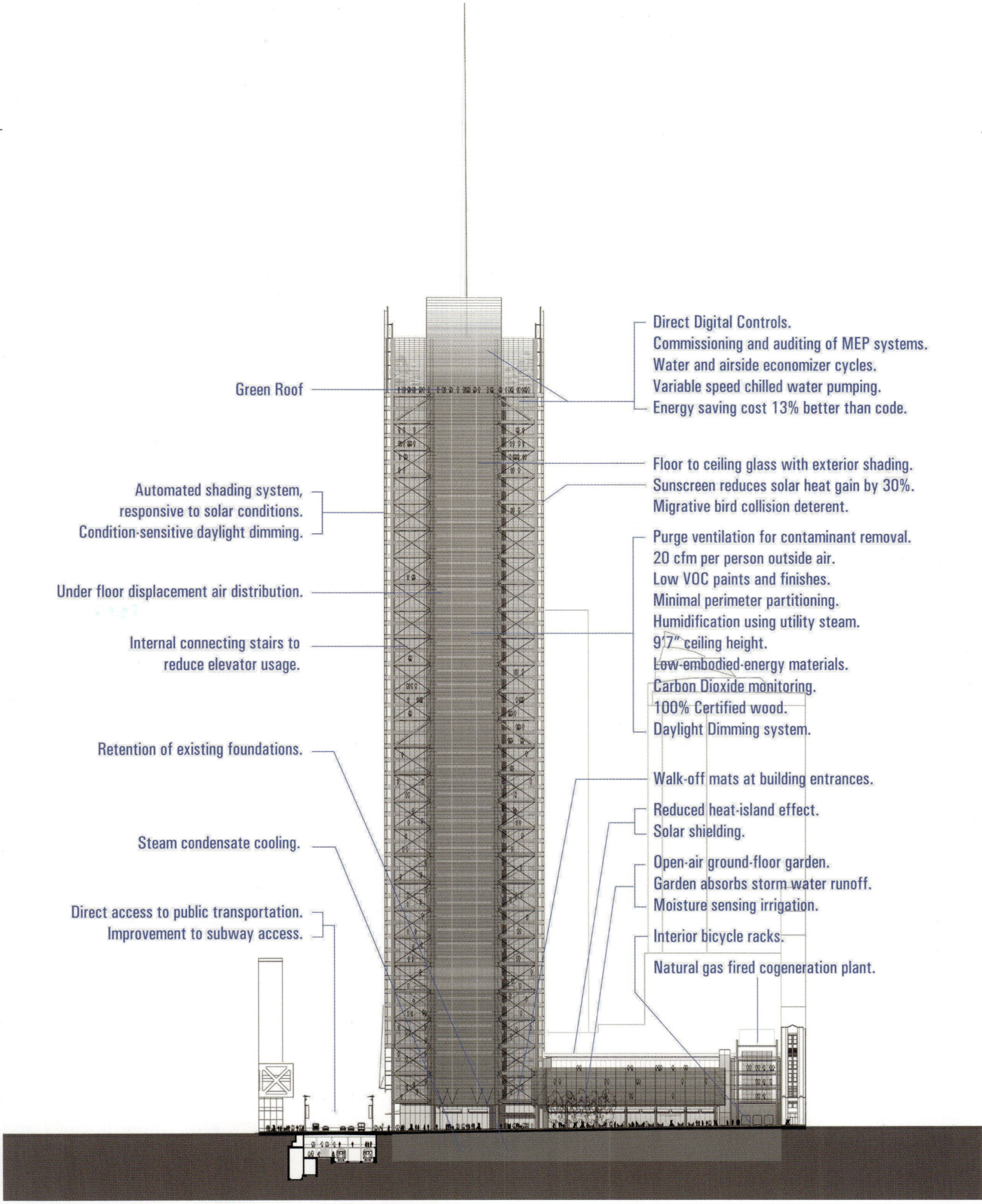

South elevation showing sustainability features

2008 NOMINEE: BEST TALL BUILDING | AMERICAS

340 ON THE PARK
CHICAGO, USA

View of the north elevation

340 on the Park stands out as a major achievement in energy and environmentally conscious design with its 62-story height and residential use. The aesthetic components of the building sought to maximize the views and to interrelate with the surrounding area. The façades each relate to the view they are facing, ensuring optimal views to the surrounding Grant Park, Lake Michigan and Chicago skyline. The striking prow point sidesteps existing views of nearby structures and creates 360º views around the site for the residents of 340 on the Park. Residents may also enjoy the winter garden located on the 25th floor. This fully accessible amenities level utilizes an entire floor and includes a two-story landscaped interior with an outdoor terrace.

The entire building design features innovative approaches to an energy efficient design; however, this was not an easy feat for the project team. Because the building is a residential condominium tower, certain standard approaches to energy conscious design did not turn out to be viable options —solar energy, wind driven turbines, and other resources typically employed in sustainable design.

> ***"The mid-level sky garden is a fantastic space, usually reserved for the exclusive world of top class hotels."***
>
> *-Antony Wood, Editor*

AWARDS JURY / EDITOR STATEMENT

A collaborative effort was needed in order to develop sustainable systems that were both cutting edge and appropriate to the design of the building. The resulting sustainable features include: Regionally manufactured and supplied core and shell materials containing recycled content, paints, adhesives, sealants and carpets that meet the LEED program requirements; a design and customization manual provided to residents to assist in the selection of interior finishes and materials; more than 75% of all roof surfaces have either a highly reflective white surface or green roof garden; 100% of all irrigation water is supplied by an internal storm water collection system to meet the already stringent ASHRAE 90.1 Energy Code standard; the building has a highly efficient envelope system. With both U and S-C coefficient values below 0.30, 82% of construction waste was recycled instead of being sent to landfills. The result is a building that looks effortless and elegant and stands as a model for the next generation of energy and environmentally conscious buildings.

View from the southeast with Grant Park in the foreground

Winter garden on 25th floor

PROJECT DETAILS

Completion Date
December 2007

Height
205 meters/672 feet

Total Area
304,800 square meters/
1,000,000 square feet

Use
Residential

Owner / Developer
Related Midwest

Architect
Solomon Cordwell Buenz

Structural Engineer
Magnusson Klemencic Associates

MEP Engineer
Cosentini Associates

Contractor
Bovis Lend Lease, Inc.

Other Consultants
Wolff Associates
Shiner + Associates, Inc.
Sieben Energy Associates

2008 NOMINEE: BEST TALL BUILDING | AMERICAS

7 WORLD TRADE CENTER
NEW YORK CITY, USA

View from the southeast; image © Vanni Archive/Ruggero Vanni

"Though not departing radically from the established commercial form for office towers, the treatment of skin and the public elements in envelope and landscaping at base establish this building as an extremely positive element where positivity is sorely needed."

-Antony Wood, Editor

AWARDS JURY / EDITOR STATEMENT

7 World Trade Center is located just north of the main site of the World Trade Center. With Lower Manhattan struggling to maintain its vitality as a leading business district before the rebuilding of the World Trade Center proper, 7 World Trade Center helped to jump-start the neighborhood's resurgence both economically and symbolically after 9/11. The footprint of the old 1980s-era building was pulled back to open up Greenwich Street, restoring the view corridor from Tribeca to the island's southern tip that had been blocked since the 70s, and creating a new public park.

The tower is notable for its innovative detailing and formal restraint. Artist James Carpenter worked with the architectural design team to create a curtain wall whose curved spandrels and reflective sills produce a unique optical effect. Both smaller and airier than its predecessor, the 52-story slender glass pylon sits atop transformer vaults which supply electricity to millions of users.

To accommodate the transformers' intense airflow requirement, the podium wrapper is designed as a breathing, stainless steel surface that gleams by day and glows at night. Behind the podium screen are blue and white LED fixtures linked to motion sensors which vary the wall's illumination in the evening as pedestrians pass by.

A special outside air intake system, which filters outdoor air (and air-conditioned air) through an activated carbon filtration system, was developed to ventilate all offices, providing a healthier indoor environment for occupants. One of the first projects accepted as part of the U.S. Green Building Council's Pilot Program for Leadership in Energy and Environmental Design (LEED) for Core and Shell Development, 7 World Trade Center is the first LEED Gold certified high-rise office building in New York City. The tower is also a model for life safety, with a reinforced concrete core, steel superstructure, and safety enhancements that exceed NYC building codes.

Standing tall as a simple yet nuanced contribution to the Manhattan cityscape, 7 World Trade Center sets a new standard for high-rise structures as it announces the Financial District's sustainable resurgence.

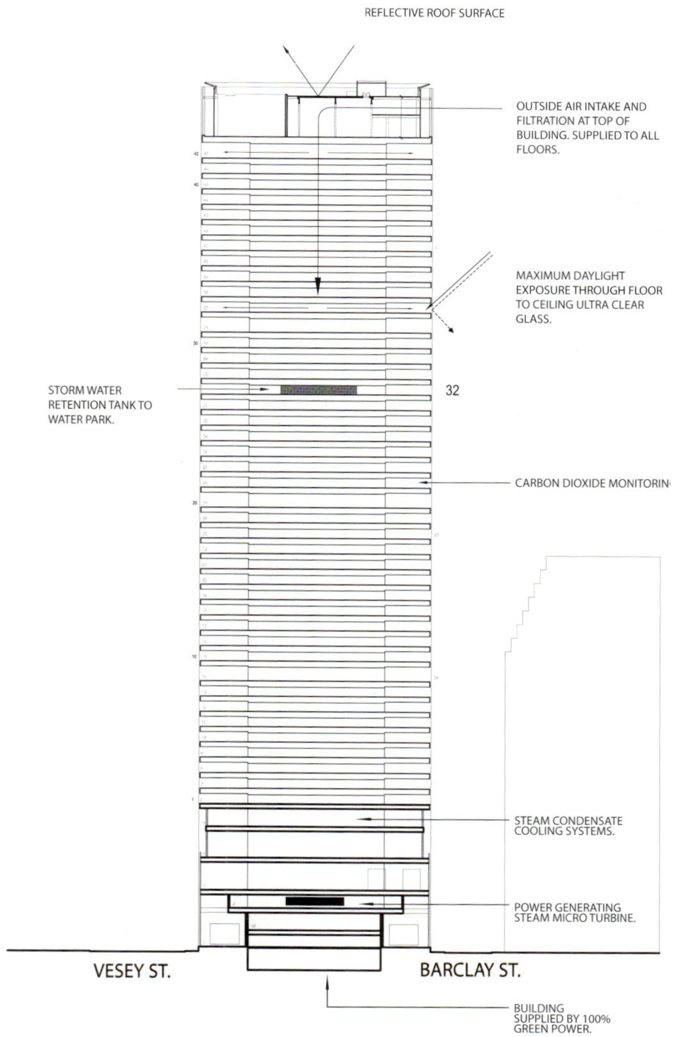

Building section

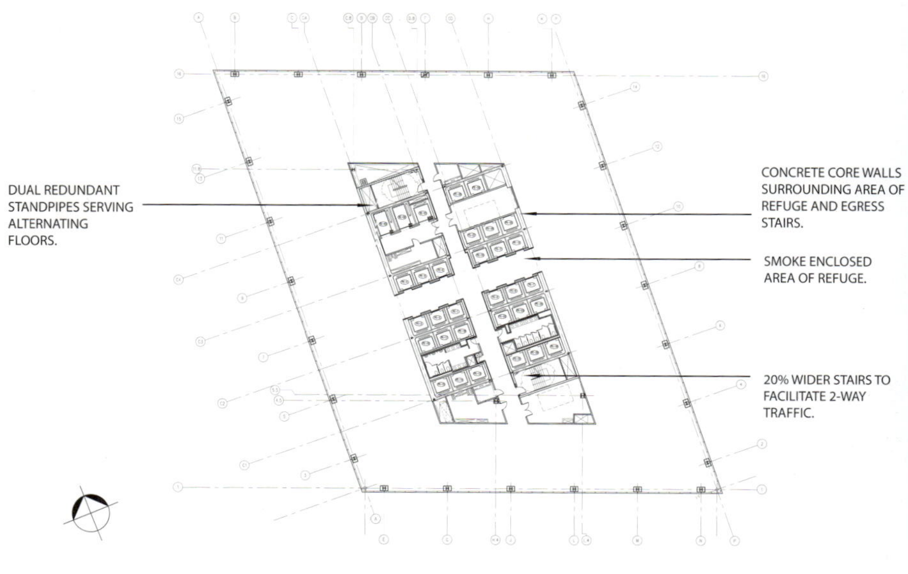

Typical floor plan

PROJECT DETAILS

Completion Date
May 2006

Height
229 meters/752 feet

Total Area
156,181 square meters/
1,681,118 square feet

Use
Office, Con Ed Power Station

Owner / Developer
Silverstein Properties, Inc.

Architect
Skidmore, Owings & Merrill LLP

Structural Engineer
WSP Cantor Seinuk

MEP Engineer
Jaros Baum and Bolles

Contractor
Tishman Construction

Other Consultants
Ken Smith Landscape Design
Philip Habib and Associates
James Carpenter Design Associates

2008 NOMINEE: BEST TALL BUILDING | AMERICAS

COMCAST CENTER
PHILADELPHIA, USA

View from southeast; image © Peter Aaron / Esto

Comcast Center is the new headquarters for the Comcast Corporation rising 57 stories, with typical floor plates ranging between 2,140 and 2,600 square meters (23,000 and 28,000 square feet). Taking the form of a faceted obelisk, the tower is clad in silvery high-performance glass with ultra-clear, low-iron glass at the building's corners and crown. The Comcast Center and its south-facing, half-acre plaza straddles the underground tracks and concourse of Suburban Station —Philadelphia's primary commuter rail gateway.

A 110-foot-high, light-flooded public winter garden connects the concourse with its shops and food hall to the tower and plaza above. The winter garden features a double-skin glass curtain wall with sunscreens and louvers which optimize daylight and views while modulating daily and seasonal thermal performance. Radiant heating, thermal extraction, and displacement ventilation combine to provide exceptional energy performance for this civic-scaled space. Three 3-story "sky atria" in the lower portion of the tower's south façade overlook the plaza and provide tenants with unique and identifiable homes. The building is expected to receive LEED-CS Certification.

View from the west; image © Peter Aaron / Esto

AWARDS JURY / EDITOR STATEMENT

"The building is a modern environmental update on some of the classic 1980's 'tower & public atrium' skyscrapers such as Philip Johnson's AT&T Building New York."

-Antony Wood, Editor

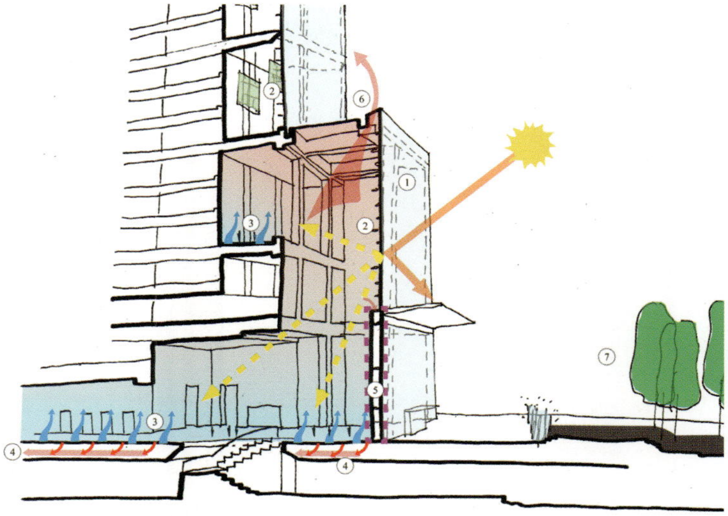

Diagram of atrium cooling scheme:
1) High performance, low-e, low-iron glass reduces solar gain by 62% while allowing 73% visible light transmission
2) Sunshades and vegetative screens reduce glare and heat gain
3) Low-velocity displacement ventilation system tempers occupied areas only
4) Hydronic, in-slab heat extraction reduces solar heat gain and improves comfort
5) Double-skin curtain wall creates thermal barrier and air return
6) Automatic heat evacuation aided by natural stack effect and controlled by heat sensors
7) Vegetation shades 30% of plaza while using rainwater retention for irrigation and fountain for evaporative cooling.

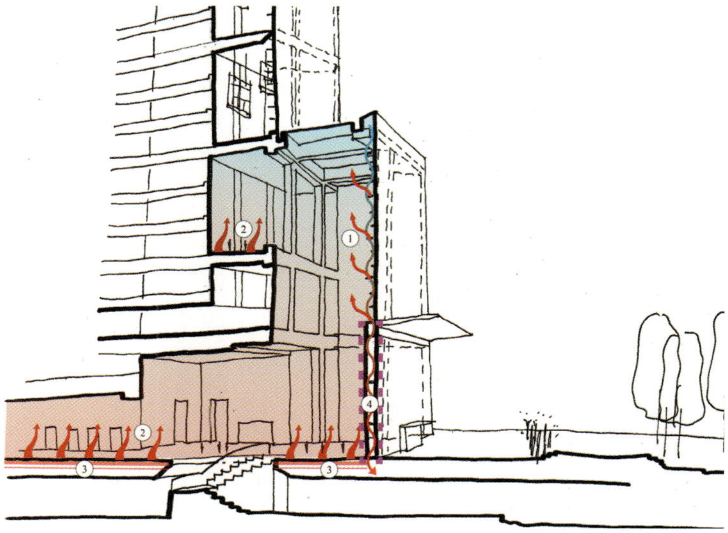

Diagram of atrium heating scheme:
1) Sunshades with internal radiant fin tubes prevent downdrafts along glass
2) Low-velocity under floor air system tempers occupied areas only
3) Dark stone floor radiates solar heat back to the space at night
4) Double-skin curtain wall creates thermal barrier and air return.

PROJECT DETAILS

Completion Date
June 8, 2008

Height
297 meters/974 feet

Total Area
130,064 square meters/
1,400,000 square feet

Use
Office

Owner / Developer
Liberty Property Trust

Architect
Robert A.M. Stern Architects, LLP/ Kendall/ Heaton Associates

Structural Engineer
Thornton Tomasetti

MEP Engineer
Paul H. Yeomans, Inc.

Contractor
L.F. Discoll Co.

Other Consultants
Atelier Ten
The Olin Partnership
Pennoni Associates
Cerami and Associates, Inc.
Gensler

2008 NOMINEE: BEST TALL BUILDING | AMERICAS

ONE RINCON HILL
SAN FRANCISCO, USA

View from the southwest

One Rincon Hill anchors the redeveloping neighborhood of Rincon Hill in San Francisco, California. Standing at 64 stories tall, the all-residential tower is located on a narrow site just eight miles from the San Andreas Fault. In response to being built in this highly seismic region, the project employs a first-of-its-kind structural system, using a nontraditional Performance Seismic Design approach (PSD), the first ever to be approved in California.

This structure features a concrete ductile core wall and four concrete outrigger columns connected by buckling restrained braces. The PSD design eliminated the need for an exterior moment frame which reduced overall cost by 8%, increased individual unit areas, allowed unobstructed views, shortened construction time, optimized material quantities, and enhanced performance.

Another first in this project is the tuned liquid-mass damper, the first to be employed in the U.S. Two 54,000-gallon tuned liquid sloshing damper tanks located at the building's top use the motion of sloshing water to counteract wind accelerations and increase occupant comfort.

Throughout the design development process the structural engineer undertook two different research projects. The first project focused on link beam connections at UCLA (with Charles Pankow Foundation and Webcor Concrete). The other studied post-tensioned slab/wall connections at UC Berkeley. The efforts improved One Rincon Hill's design, led to a modification of ACI Code 318-08, and will influence future design and construction practices.

AWARDS JURY / EDITOR STATEMENT

"This project breaks the mold in its understanding of most likely REAL performance in the event of an earthquake. In this, it lights the way for performance-based design across many disciplines: fire, safety, energy consumption, etc."

-Antony Wood, Editor

Viewed in context of Rincon Hill Neighborhood

Rendering of outrigger system, part of the Performance Seismic Design

PROJECT DETAILS

Completion Date
May 1, 2008

Height
184 meters/605 feet

Total Area
70,792 square meters/
762,000 square feet

Use
Residential

Owner / Developer
Urban West Associates

Architect
Solomon Cordwell Buenz

Structural Engineer
Magnusson Klemencic Associates

MEP Engineer
CB Engineers

Contractor
Bovis Lend Lease, Inc.

Other Consultants
Korth Sunseri Hagey Architects
Webcor Concrete

2008 **NOMINEE:**
BEST TALL BUILDING | AMERICAS

SAN FRANCISCO FEDERAL BUILDING
SAN FRANCISCO, USA

North elevation—the narrow footprint allows for wind-driven natural ventilation in the upper 13 floors; image © Cody Andresen

The San Francisco Federal Building is an office building for the U.S. Federal Government. Designed as part of the General Services Administration's Design Excellence Program, the project consists of an 18-story tower of reinforced concrete, a 4-story steel braced frame annex building, and a stand-alone pavilion for a restaurant.

To support the GSA's commitment to incorporating energy efficiency and sustainability, the team proposed a design based largely on energy reduction through natural ventilation, taking advantage of the temperate Bay area climate. Exposed thermal mass was required to supplement the natural cooling, absorbing heat from computers, people and lighting. However, downturned concrete beams would have disrupted the airflow required to cool the slab, while also blocking light. A solution of upturned beams creates an unobstructed soffit across which air flows unimpeded, promoting heat transfer. The ribbed slabs that span between the upturned beams have a wave profile varying from 12-5/8 to 4-3/4 inches to reduce mass, increase air contact surface area, improve lighting by eliminating shadows and enhance the spaces aesthetically. At night, computer-controlled windows behind the scrim open, allowing the evening air to flow through the building and cool the concrete.

A narrow 20-meter (66-foot) floor plate allows for wind-driven natural ventilation. A thinned slab at the perimeter edge maximizes window height assisting with daylight penetration. The project's tower building will consume 33% less energy compared to an office building design under the same California energy code.

AWARDS JURY / EDITOR STATEMENT

"Several cities are competing for the title of 'US Greenest City'; among them Chicago, Seattle, and San Francisco. In the high-rise sustainability stakes, this building definitely bolsters the claim for advantage west coast."

-Antony Wood, Editor

Southeast elevation featuring automated operable scrim panels that respond to the sun's position; image © Tim Griffith

Interior view of the Main Lobby's south end, featuring raking columns and natural daylight penetration; image © Tim Griffith

PROJECT DETAILS

Completion Date
June 2007

Height
71 meters/234 feet

Total Area
56,206 square meters/
605,000 square feet

Use
Federal Office

Owner / Developer
U.S. General Services Administration

Architect
Morphosis/ SmithGroup

Structural Engineer
Arup

MEP Engineer
Arup

Construction Manager
Hunt Construction Group

Other Consultants
Dick Corporation
Morganti General Contractors

BEST TALL BUILDING

WINNER

WINNER:
SHANGHAI WORLD FINANCIAL CENTER
SHANGHAI, CHINA

page 34

NOMINEE:
PARK TOWER
BEIJING
CHINA

page 56

NOMINEE:
ONE RAFFLES QUAY
SINGAPORE

page 54

NOMINEE:
THE SAIL @
MARINA BAY
SINGAPORE

page 60

NOMINEE:
NEWTON SUITES
SINGAPORE

page 48

NOMINEE:
181 WILLIAM STREET
MELBOURNE
AUSTRALIA

page 40

AWARDS 2008

ASIA & AUSTRALASIA

NOMINEE:
POLY INTERNATIONAL PLAZA
GUANGZHOU, CHINA

page 58

NOMINEE:
FUJIAN POWER HEADQUARTERS
FUZHOU, CHINA

page 44

NOMINEE:
AIG TOWER
HONG KONG
CHINA

page 42

NOMINEE:
NINA TOWER DEVELOPMENT
HONG KONG, CHINA

page 50

NOMINEE:
ONE ISLAND EAST
HONG KONG
CHINA

page 52

NOMINEE:
LUMIÈRE RESIDENCES
SYDNEY
AUSTRALIA

page 46

WINNER:

2008 BEST TALL BUILDING ASIA & AUSTRALASIA

SHANGHAI WORLD FINANCIAL CENTER

SHANGHAI, CHINA

The Shanghai World Financial Center, at 101 stories, is a symbol of commerce and culture that speaks to the city's emergence as a global capital. It is recognized by the Council on Tall Buildings and Urban Habitat as the world's tallest building in two of its four categories, height to top of roof (487 meters/1,599 feet) and highest occupied floor (474 meters/1,555 feet) (*see page 116 for more on CTBUH Height Criteria*). Located in Shanghai's Pudong District, the mixed-used SWFC is a vertical city, containing 62 office floors, conference facilities, urban retail and dining spaces, and a 174-room five-star Park Hyatt Hotel at the top—the world's highest hotel from the 79th to 93rd floors. Above the hotel, at the 94th to 100th floors, is a visitors' square and observatory, which is the highest publicly accessible built space in the world.

Shaped by the intersection of two sweeping arcs and a square prism—shapes representing ancient Chinese symbols of heaven and earth, respectively—the tower's tapering form supports programmatic efficiencies, from large floor plates at its base for offices to rectilinear floors near the top for hotel rooms (*see floor plan diagram on page 39*). Its boldest feature, the 164-foot-wide portal carved through its upper levels relieves the enormous wind pressures on the building. The project activates the ground plane through function-specific entrance volumes (e.g., hotel, office and retail) that extend from its stone-clad base. To further connect the activities of the building to the city, the retail volume is oriented toward a public park planned for an adjacent site.

Optimizing form and function was paramount to the design, integrating the structure, mechanical systems, and exterior envelope in a modular system that repeats every 13 floors to facilitate the fabrication and installation of components, and, in turn, reduce construction time, material waste, and structural inefficiencies. The purity of the tower's design belies the inherent complexity of the various building systems within, and is readily adaptable to the changing programmatic requirements that often arise during the long timeline of such a large project, as well as to the changing needs of building users.

The project was put on hold in 1995 after the completion of the foundations. When revived in 1999 the height and base

AWARDS JURY/EDITOR STATEMENT

"This is a building that inspires an impression of its place. The building structure is nothing short of genius."

-Tim Johnson, NBBJ

View from the north with Jin Mao Tower standing across the street at right

PROJECT DETAILS

Completion Date
August 2008

Height
492 meters/1,614 feet

Total Area
377,300 square meters/
4,061,223 square feet

Use
Office, Hotel, Retail,
Conference Facilities,
Observation Deck

Owner / Developer
Mori Building Company Ltd.

Architect
Kohn Pedersen Fox
Associates PC/
Irie Miyake Architects and
Engineers

Structural Engineer
Leslie E. Robertson
Associates, R.L.L.P.

MEP Engineer
Kenchiku Setubi Sekkei
Kenkyusho

Contractor
China State Construction
Engineering Corporation/
Shanghai Construction
General Company

SHANGHAI WORLD FINANCIAL CENTER

SHANGHAI, CHINA

Viewed from the southeast

dimensions were both increased from the original design. Reinforcing the existing piles to accommodate these changes would have been possible but costly. The new, taller structure would not only have to be made lighter, but would need to resist higher wind loads, which increase exponentially with height.

The project's structural engineer developed a new system, employing composite mega-columns, diagonal mega-braces, steel outriggers, belt trusses, and core wall trusses (*see right*), the pile loads were redistributed to accept increased lateral loads from wind and earthquake. The stiffness of the lateral force-resisting system of the perimeter wall was increased, and as such, the original design for the perimeter framing was abandoned in favor of a diagonal-braced frame with added outrigger trusses coupled to the columns of the mega-structure. This enabled the weight of the building to be reduced by more than 10% and resulted in a reduced cost for the structure, provided for speedier construction, and significantly reduced the material that went into the building and thus made the building even more environmentally friendly.

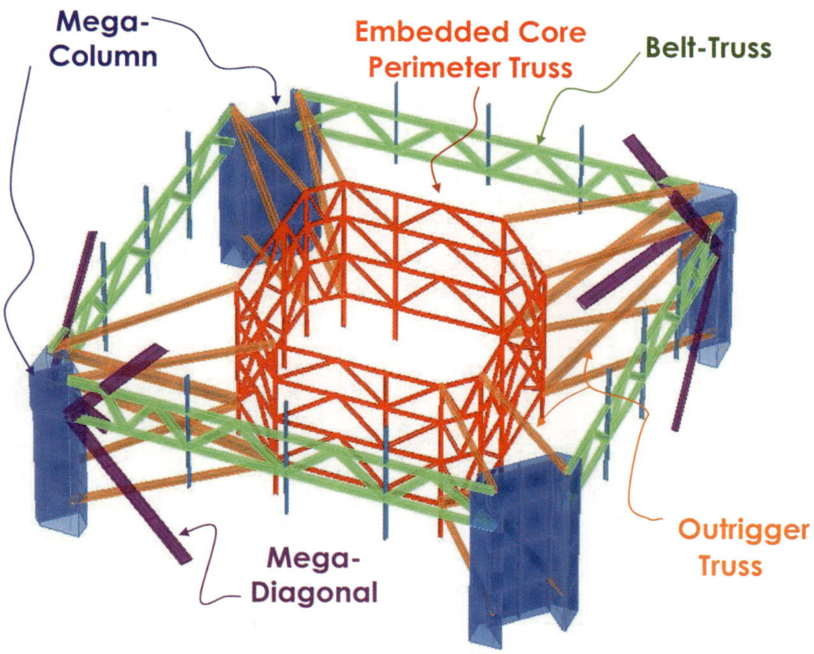

Structural diagram showing how the various components of structure work together

Construction photo showing exposed structural elements before cladding

Model showing the building's structure system of mega-columns, diagonals and belt trusses along with concrete core walls and outrigger trusses

SHANGHAI WORLD FINANCIAL CENTER

SHANGHAI, CHINA

AWARDS JURY / EDITOR STATEMENT

> ***"The Shanghai World Financial Center now completes the two-decade dialogue with its Jin Mao neighbor, both different but confident interpretations of the Chinese skyscraper."***
>
> *-Antony Wood, Editor*

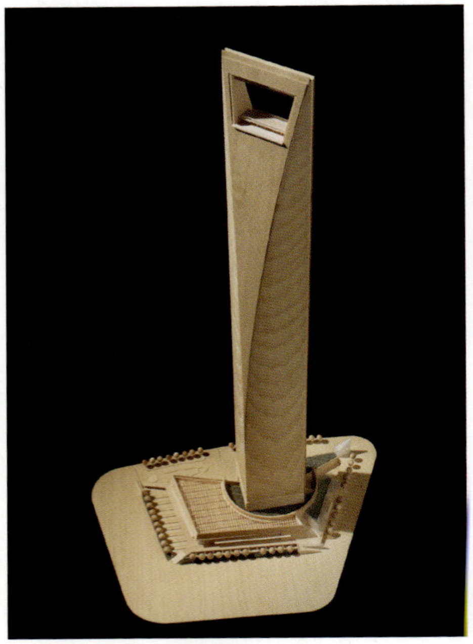

Architectural model of tower

Awards Jury Statement:

Asia is a region with half the world's population, and currently undergoing unprecedented development. With the greatest rate of migration in the history of humankind from rural to urban areas there is an explosion of high rise development everywhere. Density is key, as well as the craving for a recognizable and modern identity. High rise buildings play an essential role in achieving these desires. The Asia region had a significant amount of very high quality submissions for a CTBUH award this year and fostered a wonderful debate about selecting the best. In the end the Shanghai World Financial Center was selected the winner. First of all—this structure is a lesson in endurance having been started in the 1990's and now just being finished. As one juror commented "it takes six miracles for anything great to happen". This project has become the icon of Shanghai and potentially China. Its simple and clear form is dramatic at all scales through a connection metaphorically of the earth and sky. Its structural design is revolutionary. The building is profound —it speaks to where tall building design is today.

Viewed against the Shanghai skyline

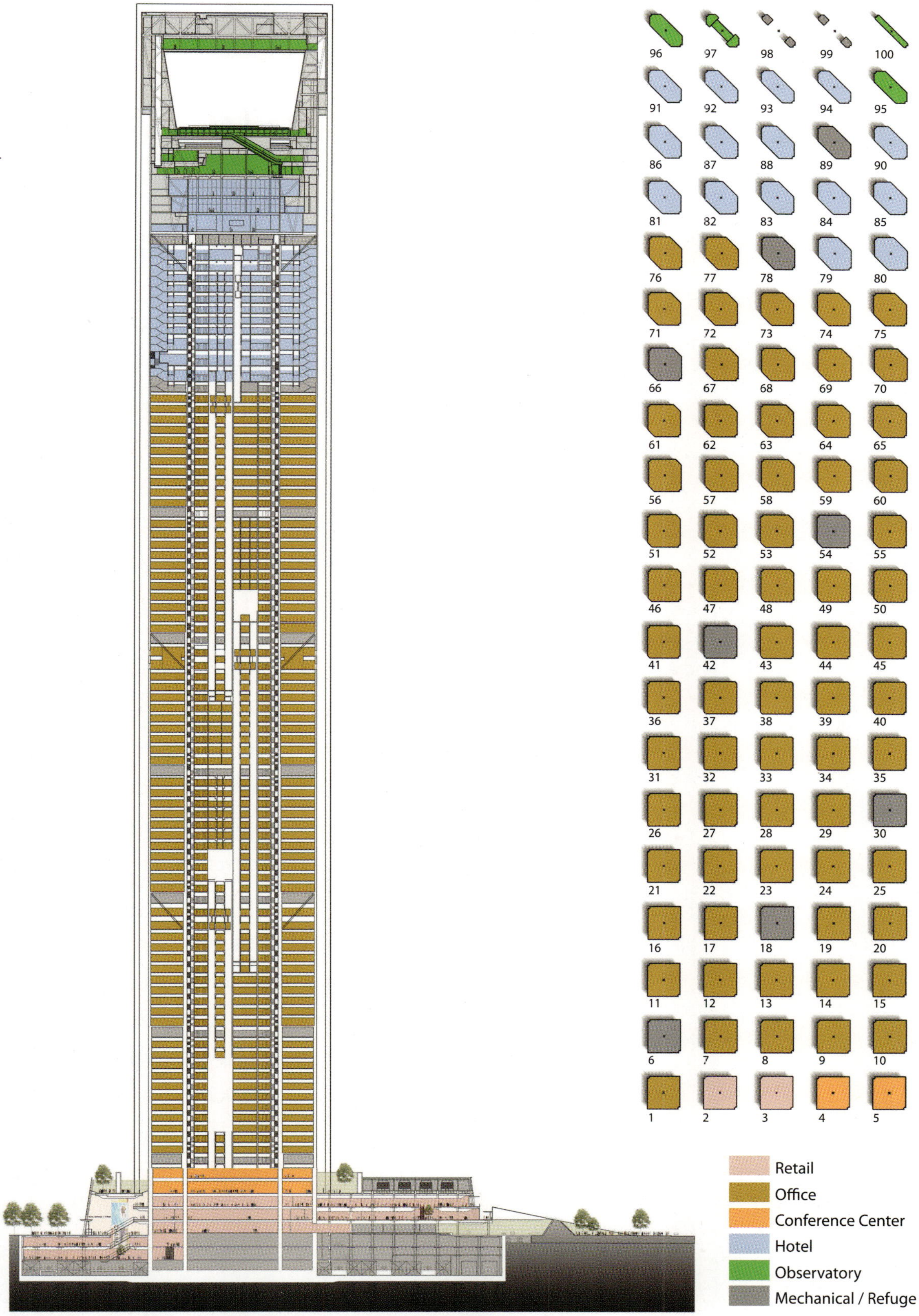

Color coded section and series of floor plans showing the progressive change in form

2008 NOMINEE: BEST TALL BUILDING | ASIA & AUSTRALASIA

181 WILLIAM STREET
MELBOURNE, AUSTRALIA

View of tower from corner of Bourke and William Streets

181 William Street is a new major corporate presence in the high ground of the west end commercial precinct of Melbourne. The entire project is designed on a holistic basis to integrate the façade with all services systems to achieve high quality environmental performance. During the course of a comprehensive design process, all services systems were integrated to ensure high quality environmental performance, flexibility and long life.

The tower façade is visually arresting in the use of colors in the vision and spandrel glass, creating a balanced geometric pattern that captures a sense of movement across the façade. The russet color pallet was selected to reflect a richer and more distinct corporate presence. The range of seven colors to the spandrel and vision glass has been distributed in an ordered manner that has been inspired by the dot painting technique of Australian indigenous art. The white columns are metal clad with the lower sections expressed in polished stone. The elevated podium is clad in zinc with expressed joints and flush glazing to match. The north and south inset façades are in clear flush glazing.

The tower is designed to achieve maximum natural light, views and workplace flexibility. The project has achieved a certification rating of Five Green Stars and 4.5 Star ABGR. These rating systems are administered by the Green Building Council of Australia and subject to rigorous audit and monitoring. These assessments cover all elements of sustainable performance, including the consideration of embodied energy, recycled construction materials, indoor air quality and VOC analysis, and ongoing environmental performance. This is equivalent to a LEED Gold rating.

AWARDS JURY / EDITOR STATEMENT

"The move to integrate services and façade is an interesting one with a multitude of possibilities for the future: does this indicate the next, more-intensive stage of the evolution of the double-skin façade?"

-Antony Wood, Editor

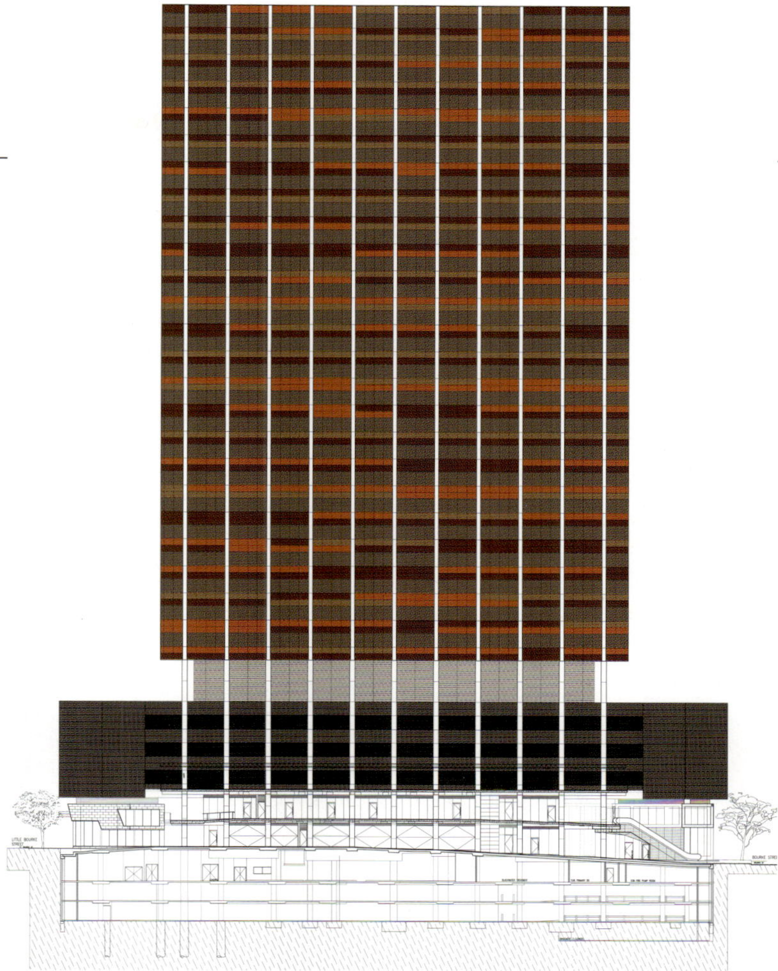

West elevation showing distribution of the seven colored glass panels

Rendering of main lobby

PROJECT DETAILS

Completion Date
May 2008

Height
123 meters/404 feet

Total Area
49,601 square meters/
533,901 square feet

Use
Office, Retail

Owner / Developer
CBus Property

Architect
Bates Smart + SJB

Structural Engineer
Winward Structures

MEP Engineer
Norman Disney & Young

Contractor
Multiplex Constructions Pty Ltd.

Other Consultants
Urbis JHD Pty Ltd
McKenzie Group Consulting
Rider Levett Bucknall Limited
Arup

2008 NOMINEE:
BEST TALL BUILDING | ASIA & AUSTRALASIA

AIG TOWER
HONG KONG, CHINA

View from the southeast

AIG Tower is located in the Central District of Hong Kong on a site that was most recently home to the former Furama Hotel. Developed as a joint venture between AIG, CapitaLand, and Lai Sun Development, the building holds further distinction as the site where AIG opened its first office in the territory in 1930. The architectural design draws upon the visual imagery of Chinese junks that were vital to the city's economic growth. The building's form, tapered at its base and apex on three sides of the building, evokes the sails of the vessels which once crossed Victoria Harbor.

Large floor plates provide modern, flexible space and reduce the number of stories needed to meet Owner requirements. Structurally, the lateral load resisting system consists of a reinforced concrete core coupled with exterior steel columns through two-story steel trusses at the mid-rise refuge floor and the third floor mechanical level. Serving dual functions as outrigger and transfer trusses, this resulted in an efficient structural system with a high degree of robustness, while providing flexibility in meeting architectural and programming requirements.

> ***"The re-use of foundations from a previous project on the site is a sensible strategy, especially in today's increasingly carbon-challenged and credit-crunched world."***
>
> *-Antony Wood, Editor*

AWARDS JURY / EDITOR STATEMENT

Additional program considerations include the reuse of foundations from the Furama Hotel, which minimized the impact to neighboring structures; and adhering to stringent vibration requirements due to the proximity of the Mass Transit Railway, which prevented the use of additional pilings. Furthermore, the fewer floors of building core, lift shafts, and stairs effectively reduced the amount of construction materials, resulting in a more environmentally friendly addition to the Hong Kong skyline. These structural contributions to the project were integral to the "Excellent" rating issued by the Hong Kong Building Environmental Assessment Method (HK-BEAM).

Finally, the striking design of the associated footbridge advances the state of the engineer's art. The new footbridge links the AIG Tower to the Hong Kong elevated walkway network and enhances convenience for pedestrians. Spanning 45 meters (150 feet), the partially enclosed bridge is formed from an asymmetrical box truss, S-shaped in plan. The truss is supported by pipe diagonals of diminishing sizes which architecturally express the flow of forces.

Night view looking up at the northeast corner

Rendering of the footbridge

PROJECT DETAILS

Completion Date
March 2006

Height
185 meters/607 feet

Total Area
39,000 square meters/
419,800 square feet

Use
Office

Owner / Developer
AIG/ CapitaLand/ Lai Sun Development

Architect
Skidmore, Owings & Merrill LLP

Structural Engineer
Leslie E. Robertson Associates, R.L.L.P.

MEP Engineer
WSP Flack + Kurtz

Contractor
Hip Hing Construction Co. Ltd.

Other Consultants
Van Deusen & Associates
MVA (HK) Ltd.
Davis Langdon & Seah (HK)

2008 **NOMINEE:**
BEST TALL BUILDING | ASIA & AUSTRALASIA

FUJIAN PROVINCIAL ELECTRIC AND POWER COMPANY HEADQUARTERS
FUZHOU, CHINA

View from the east

The Fujian Provincial Electric and Power Company Headquarters embodies the high-tech characteristics and modernized features of an electric power company by providing a first-rate working environment that is efficient, advanced, healthy and comfortable, as well as an icon for the city of Fuzhou. The sculptural form of the main tower adopts a mono-axially symmetric rhombus and is set at the site's northern-most position to maintain prominence in the developing skyline, and maximizes site area to be used for public green space. The design concept creates a unique, yet culturally relevant, building.

Energy conduction is the main thematic motif. The crown incorporates a viewing space within the microwave tower. The building's sail-like form pays tribute to the seafaring people of Fuzhou. The building uses a unique revolving aluminum ventilation assembly instead of conventional operable windows to provide a natural, easy to maintain, and cost-effective ventilation system. Additionally, this vertically integrated green system makes it a truly healthy working environment.

View of base nearing completion

PROJECT DETAILS

Completion Date
July 2007

Height
132 meters/434 feet

Total Area
63,278 square meters/
681,118 square feet

Use
Office, Data center

Owner / Developer
Fujian Electric Power Co.

Architect
MulvannyG2

Structural Engineer
Northeast China
Architectural and Design
Institute

MEP Engineer
Northeast China
Architectural and Design
Institute

Contractor
Northeast China
Architectural and Design
Institute

AWARDS JURY / EDITOR STATEMENT

"The building is interesting, in reflecting the growing desire of companies in the power-generation business to portray their sustainability credentials."

-Antony Wood, Editor

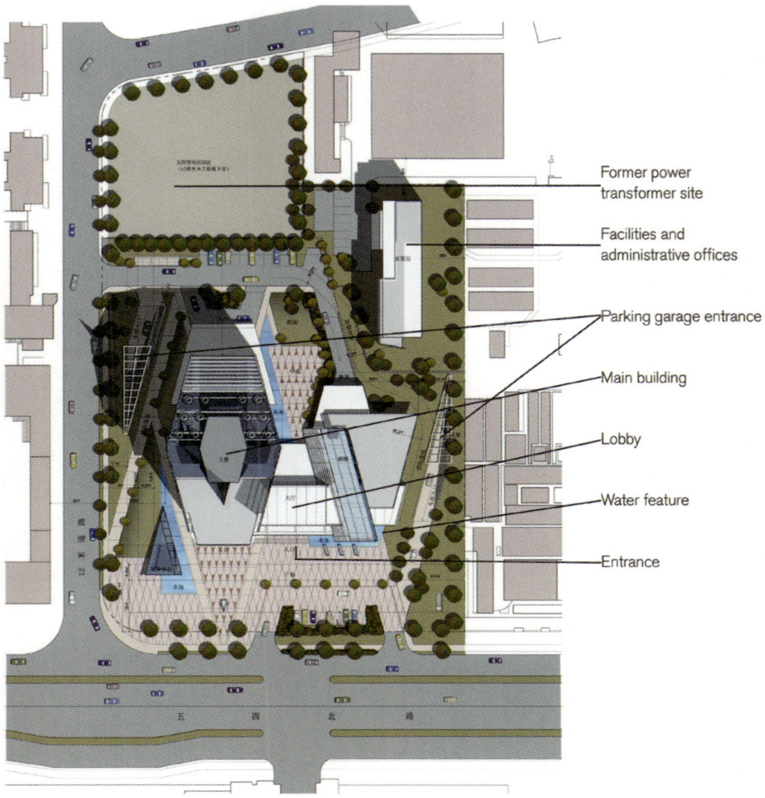

Site plan

2008 NOMINEE: BEST TALL BUILDING | ASIA & AUSTRALASIA

LUMIÈRE RESIDENCES
SYDNEY, AUSTRALIA

View from the north; image © Tom Evangelidis/ Frasers Property

Lumière Residences is the main 56-story residential tower within the Regent Place mixed-use development at the heart of Sydney. Eight slender volumes supported by a central core provide a diverse village of 456 apartments floating above a five-level podium, which reflects the scale of adjacent historic buildings and engages with the city center through mixed retail, commercial and leisure spaces. As a point of reference for the city skyline, the tower becomes increasingly transparent as it rises into the sky, its floating effect accentuated by black glazed necking of the façade above the podium, where lush landscaped terraces mingle visually with street tree-tops.

Projecting fins to the edge of each sleek glass and metal volume enhance their strong vertical form, while deep recesses into the heart of the tower allow daylight, cross ventilation and views from each lobby. Balconies integrated within the metal and glass façade create loggia for year-round use while preserving a seamless appearance in harmony with neighboring buildings. Internal folding screen doors, combined

> ***"The use of double-skin façade and folding internal glazed screens introduces a blurring of interior and exterior space rarely seen in high-rise buildings."***
>
> *-Antony Wood, Editor*

AWARDS JURY / EDITOR STATEMENT

with sliding and awning windows, adapt to changes in climate and preferred use, while lower dark glazing maximizes privacy and outlook. The utilization of double glazing and water-efficient fixtures contributes to this sustainable strategy.

Typical upper level floor plan

Interior / exterior space created by retracting the folding screen doors; image © Tom Evangelidis/ Frasers Property

PROJECT DETAILS

Completion Date
November 2007

Height
150 meters/492 feet

Total Area
54,050 square meters/
581,789 square feet

Use
Residential

Owner / Developer
Frasers Property

Architect
Foster + Partners/
PTW Architects

Structural Engineer
Taylor Thomson Whitting/
Robert Bird & Partners

MEP & Façade Engineer
Connell Mott MacDonald

Contractor
Multiplex Constructions Pty Ltd.

Other Consultants
Tract Consultants

2008 NOMINEE:
BEST TALL BUILDING | ASIA & AUSTRALASIA

NEWTON SUITES
SINGAPORE

Aerial view in context

Newton Suites, a 36-story development, is a study in environmental solutions to tropical high-rise living. The design integrates several sustainable devices into a contemporary architectural composition, creating a sustainable, contemporary addition to the city skyline. The building sits at the edge of a high-rise zone and fronts a height-controlled area that affords expansive views of the central nature reserves; a rare luxury in densely built Singapore.

The exterior of the tower uses sun shading elements, patterned planes of textured panels and protruding balconies to create a façade that is functional yet expressive. The horizontal, metal expanded mesh sun shading screens the strong tropical sunlight. The angled mesh reduces solar radiation while permitting visual connection to the ground. The angled expanded mesh changes appearance with viewpoint, appearing anywhere between solid and transparent. This, combined with the cast shadows and interference patterns between the shadows and the mesh, gives the building a constantly shifting, blurred appearance depending on angle and time of day. The layers of

"With sustainable design approaches comes the opportunity for a new asethetic in tall buildings based on a natural pallette. This project will be judged by history as pointing the way forward."

-Antony Wood, Editor

AWARDS JURY / EDITOR STATEMENT

PROJECT DETAILS

sun shades also change the reading of the projections of the bay windows, a standard feature of Singapore apartments due to their contribution to developer profit and prescriptive regulations, embedding them in the language of the building.

Landscape is used as a material—rooftop planting, sky gardens and green walls are incorporated into the design. Creeper screens are applied to otherwise blank walls to create visual delight, absorb sunlight and carbon and create oxygen in the dense urban environment. Most available horizontal and vertical surfaces are landscaped; creating an area of landscaping that is 130% (110% planted) of the total site. Trees cover the car park, project from the sky gardens at every four levels and crown the building at the penthouse roof decks. The above-ground car park uses far less energy than an underground car park and is fully enclosed with creepers, absorbing exhaust emissions. The car park roof houses a substantial clubhouse with gym, steam room, party areas and 25-meter (82-foot) swimming pool with a glass overflow edge.

Vertical greening of the façade with sky gardens and a vertical planted wall

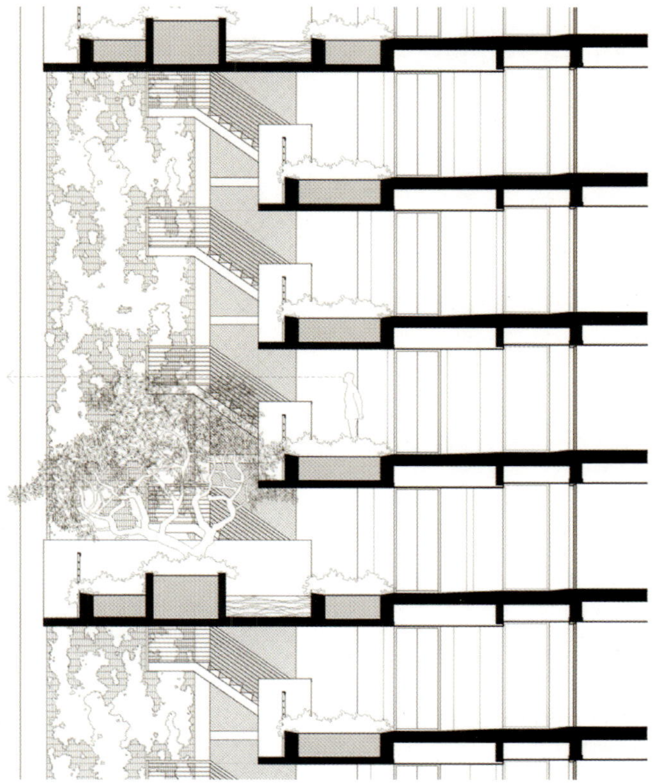
Partial section showing sky garden

Completion Date
June 30, 2007

Height
120 meters/394 feet

Total Area
11,835 square meters/
127,390 square feet

Use
Residential

Owner / Developer
UOL Group Ltd.

Architect
WOHA Architects Pte Ltd.

Structural Engineer
LBW Consultants

MEP Engineer
Lincolne Scott Ng Pte Ltd.

Contractor
Kajima Overseas Asia Pte Ltd.

Other Consultants
Cicada Pte Ltd.

2008 NOMINEE:
BEST TALL BUILDING | ASIA & AUSTRALASIA

NINA TOWER DEVELOPMENT
HONG KONG, CHINA

Aerial view of development

The Nina Tower Development consists of two towers; one stands at 80 stories and the other rises to 42 stories at 170 meters (558 feet). The taller of the two buildings, called Teddy Tower, was the tallest building to be completed during 2006. This building was originally intended to stand 200 meters (656 feet) higher than it does currently and would have been the tallest building in the world at the time of concept planning. However, the height was restricted by the Government due to the proximity of the tower to the nearby Chek Lap Kok Airport.

To stand against strong typhoons in Hong Kong, the Teddy Tower is stabilized by two levels of steel outriggers. It acts together with a reinforced concrete core wall and perimeter composite columns. A post-tensioned concrete flat slab with perimeter beam makes up the floor structure. Tailor-made table system formwork for the floor, and climb form for the core wall were used and thus sped up the construction cycle.

A shopping mall and exhibition spaces occupy the lower stories of the development, which is topped by 30 stories of offices in the Teddy Tower. The smaller tower (dubbed the Nina Tower) and the top 40 stories of the Teddy Tower house a five-star hotel. The two towers are connected by a Skybridge at the top of the shorter building. The Skybridge linking the two towers is curved in plan and the structure comprises two 3-meter (10-foot)-deep triangular space trusses.

The Nina Tower Development plays an essential role in continuing to bring Hong Kong into a new era of modern architecture and development. It also serves to provide an exciting, state of the art destination for visitors and at the same time offers a magnificent office and hotel facility for local businesses.

AWARDS JURY / EDITOR STATEMENT

"As we make a push for ever taller, denser cities, connections between buildings in the form of skybridges such as that employed here at Nina need to become an urban reality."

-Antony Wood, Editor

View of Nina Tower

Skybridge connecting the two towers

PROJECT DETAILS

Completion Date
September 2006

Height
319 meters/1,046 feet

Total Area
190,000 square meters/ 2,045,143 square feet

Use
Office, Hotel, Retail

Owner / Developer
Chinachem Group

Architect
Casa Design International Limited

Structural Engineer
Arup

MEP Engineer
Arup

Contractor
Forecast Investment Limited

Other Consultants
Arup Facade Engineering

2008 NOMINEE: BEST TALL BUILDING | ASIA & AUSTRALASIA

ONE ISLAND EAST
HONG KONG, CHINA

View from east

One Island East is the latest office development at TaiKoo Place, Quarry Bay, Hong Kong. It comprises 59 office floors with a typical floor plate of 2,300 square meters (24,760 square feet). One Island East is the landmark of the district, and the center piece of this office campus.

The basic form of the building is a square plan with rounded corners and a central core. The two corners facing north and south open up at the top floors to address the Harbor view. At the base, the two corners facing east and west open up to address the open space. The edges of the four façades sail beyond, creating a floating effect and giving lightness to the building. Lighting features are incorporated into the edges of the four façades to enhance the floating effect. Architectural fins are introduced on the façade in a staggered pattern to add texture and scale.

Without a podium structure, the tower sits freely in front of a large landscaped open space to the east. The canopy at the porte cohere is specially designed as a piece of sculpture. The landscaped area beyond is designed to have platforms at different levels incorporating water features. This urban landscape gives an appropriate scale as a forecourt to the building and provides a leisure space for both the enjoyment of office workers and for the neighborhood.

The upper main lobby is connected to the rest of the office complex at TaiKoo Place via a bridge at the northwest corner. Through a well-established system of link bridges at the existing podium level, pedestrians can gain convenient access to the MTR Quarry Bay Station. The One Island East Building not only energizes the whole of TaiKoo Place, it also plays a leading role in transforming the neighborhood into a better environment.

AWARDS JURY / EDITOR STATEMENT

"The articulated edges of the building, and the staggered fins, give a subtle sculptural quality to One Island East."

-Antony Wood, Editor

View up at the tower's base

Glass footbridge connecting to existing elevated walkway system in TaiKoo Place

PROJECT DETAILS

Completion Date
March 2008

Height
298 meters/978 feet

Total Area
140,000 square meters/
1,506,947 square feet

Use
Office

Owner / Developer
Swire Properties Ltd.

Architect
Wong & Ouyang (HK) Ltd.

Structural Engineer
Arup

MEP Engineer
Meinhardt Pte Ltd.

Other Consultants
Rider Levett Bucknall
Limited

2008 NOMINEE: BEST TALL BUILDING | ASIA & AUSTRALASIA

ONE RAFFLES QUAY
SINGAPORE

Night view from northwest

One Raffles Quay (ORQ) was designed and constructed to overcome several unprecedented engineering challenges posed by the presence of existing twin subway tunnels running below the site and the highly variable and soft ground conditions.

At the planning stage, the massing of the North Tower was rotated relative to the site to allow the shortest structural transfer span and most balanced building load transmission to bridge across the subway tunnels. To transfer the 50-story tower loads, conventional transfer structures would have been too deep and would have obstructed the ground level lobby. To overcome this challenge the transfer of interior tower loads is achieved through the central box core acting as a transfer structure—probably a world first for a building of such height and complexity. At the perimeter of the tower, loads are transferred through visible and iconic mega steel trusses and raking columns in the main lobby.

The presence of up to 35-meter (115-foot)-deep layers of soft marine clay below the site presented a significant challenge to the construction of the basements since the slightest excessive ground movement could impair the operation and safety of the subway system. The design solution comprised a rigid perimeter diaphragm wall-retention system coupled with two rows of Shielding Walls (protective non-load bearing diaphragm walls) adjacent to the subway tunnels. This—along with comprehensive real time monitoring of ground, wall, struts and subway tunnel movements—allowed excavation and substructure construction with less than 15mm tunnel movement and without any adverse impact on the subway.

AWARDS JURY / EDITOR STATEMENT

"In a quiet, unpretentious way, Singapore leads the way in many aspects of tall building design, especially in its brave social and technological experiments. One Raffles Quay continues that tradition."

-Antony Wood, Editor

North tower lobby with visible transfer system

Underground pedestrian mall

PROJECT DETAILS

Completion Date
October 26, 2006

Height
245 meters/804 feet

Total Area
214,000 square meters/
2,303,477 square feet

Use
Office

Owner / Developer
One Raffles Quay Pte Ltd.

Architect
Kohn Pederson Fox Associates PC/ Architects 61 Pte Ltd.

Structural Engineer
Meinhardt Pte Ltd.

MEP Engineer
Meinhardt Pte Ltd.

Contractor
Obayashi Co.

2008 NOMINEE: BEST TALL BUILDING | ASIA & AUSTRALASIA

PARK TOWER AT BEIJING YINTAI CENTRE
BEIJING, CHINA

Night view of entire Yintai Centre

Beijing Yintai Centre, with architectural details that suggest ancient Chinese architecture in a contemporary way, is a comprehensive mixed-use complex located in Beijing's Central Business District. The three towers, square in form, are studies in simple, straightforward design. Park Tower, the focal point of the project, is the residential tower. Flanked by twin office towers, it features a five-star Park Hyatt hotel.

Topping Park Tower is a majestic cube that recalls a Chinese lantern. Express elevators sweep guests to the top of the building where the hotel registration area is located within this lantern. The cube is a stately symbol of Chinese culture by day, and a glowing landmark at night.

On top of the podium, a landscaped roof garden ties Park Tower together with the rest of Beijing Yintai Centre. Rather than attempting to copy a traditional Chinese garden, elements of historic Chinese gardens are incorporated in a symbolic way. The cool greenery and the sound of water create a welcome antithesis to the urban surroundings while also highlighting the forms of the architecture.

"This project definitely responds to its local context and exhibits something of the orient, with its rooftop 'Emperor's Box' of jewels shining out over the city."

-Antony Wood, Editor

AWARDS JURY / EDITOR STATEMENT

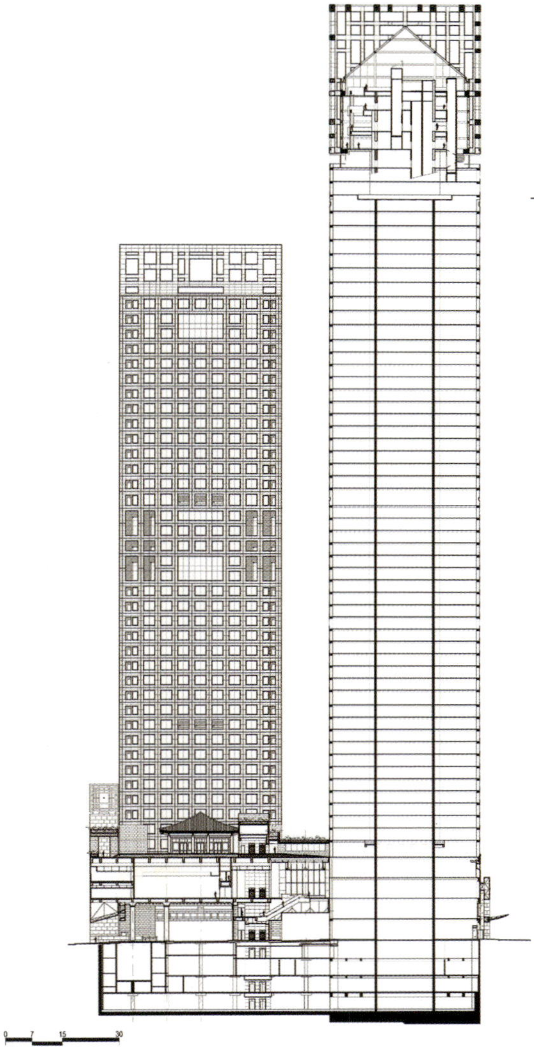

Section through Park Tower

Rendering of roof garden

Rendering of "Lantern" detail in night skyline

PROJECT DETAILS

Completion Date
August 2008

Height
250 meters/820 feet

Total Area
118,519 square meters/
1,275,728 square feet

Use
Residential, Hotel, Office, Retail

Owner / Developer
Beijing Yintai Property Co. Ltd.

Architect
John Portman & Associates

Structural Engineer
John Portman & Associates/
Le Messurier

MEP Engineer
Newcomb & Boyd / Parsons Brinckerhoff Ltd.

Contractor
Beijing Urban Construction Group (BUCG)

Other Consultants
Arnold Associates
Placemedia Landscape Architects

2008 NOMINEE: BEST TALL BUILDING | ASIA & AUSTRALASIA

POLY INTERNATIONAL PLAZA
GUANGZHOU, CHINA

Night view showing exposed lattice structure; image © Tim Griffith

Poly International Plaza is an innovative office and trade center development that consists of two towers, each coupled with a low-rise podium that accommodates exhibition and retail uses; the below-grade spaces include an exhibition hall and trade center. The two 35-story offset core towers, which include a thin bar of office space, maximize light transmission to their interiors. On the north, the towers' forms are defined by floor-to-ceiling glass, shaded by vertical fins, and on the south they expose their structural framework. The towers advocate sustainable design through their use of natural light; natural ventilation; floor-by-floor mechanical systems; under-floor air distribution; shaded outdoor space; and green roofs.

Thanks to the towers' innovative structural spine, and double lattice brace, the offset cores are light and transparent, and allow a high degree of openness to northern views; the glass elevators, staircases, restrooms, and terraces are all light-filled spaces. These terraces, which also act as emergency refuges, are the towers' social hubs thanks to their three-story floating lounges and meeting room spaces.

AWARDS JURY / EDITOR STATEMENT

"This building has many excellent attributes: a well-proportioned floor space with access to natural light, external structure providing shade as well as support, and a lifted up 'public' ground floor plane."

-Antony Wood, Editor

Lattice structure visible in both towers; image © Tim Griffith

View at ground level; image © Tim Griffith

PROJECT DETAILS

Completion Date
March 3, 2007

Height
165 meters/541 feet

Total Area
196,000 square meters/
2,108,960 square feet

Use
Office, Retail

Owner / Developer
Poly Real Estate Group Co. Ltd.

Architect
Skidmore, Owings, & Merrill LLP

Structural Engineer
Skidmore, Owings & Merrill LLP

MEP Engineer
WSP Flack + Kurtz

Contractor
Guangzhou No. 2 Construction & Engineering Co., Ltd.

Other Consultants
Guangzhou Design Institute
SWA Group
Guangzhou Gardens Planning & Building Design Institute

THE SAIL @ MARINA BAY

SINGAPORE

Overall view

The Sail @ Marina Bay takes its inspiration from the wind, water and sun. The design resembles a sculpture created by the hands of nature, with two soaring towers—split as if by water—at a sculpted base. The mixed-use building will launch the central business district's transition to a "Live/Work/Play" philosophy by providing 1,111 apartment units, restaurants, health clubs, recreation decks with pools and tennis courts, and parking.

This tower is poised to become a new destination within the rich and diverse fabric of Singapore, while setting a benchmark for the future buildings along Marina Bay. Over 65% of the apartments have panoramic views of the Marina Bay and the central city's cultural, entertainment, and retail areas. The five "6-star" penthouses are the highest elevation for any residence in Singapore, with breathtaking views of the sea and islands beyond. Many units feature "borrowed light" bedrooms in which sliding glass doors separate the bedroom from the main space while still allowing light into the bedroom area. The project includes amenities such as a hotel-style concierge service, a spa, and an executive club lounge with dining facilities.

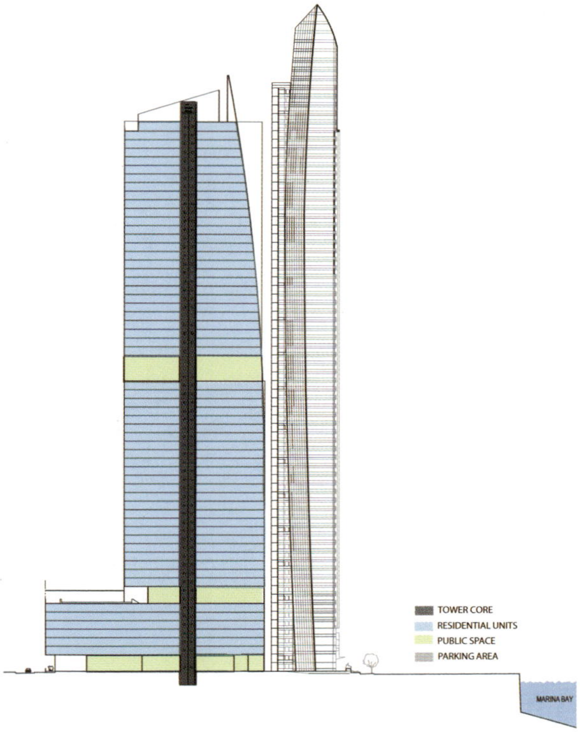

Tower two longitudinal section with tower one in elevation beyond

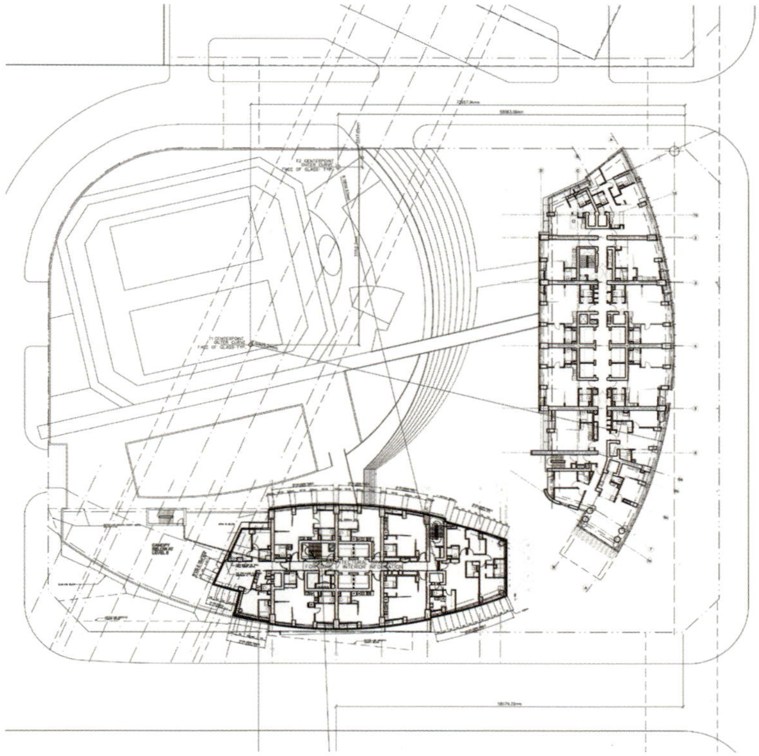

Site plan

AWARDS JURY / EDITOR STATEMENT

"The project has a great sculptural quality, clearly relating to the building's title and its water front location as part of Marina Bay."

-Antony Wood, Editor

Interior view of "borrowed light" one bedroom unit

PROJECT DETAILS

Completion Date
August 2008

Height
245 meters/804 feet

Total Area
115,000 square meters/
1,200,000 square feet

Use
Residential

Owner / Developer
City Development Ltd./ AIG

Architect
NBBJ/ Team Design

Structural Engineer
Precast Design Consultants/
Meinhardt Pte Ltd.

MEP Engineer
Meinhardt Pte Ltd.

Contractor
Dragages Singapore -
Bouygues Construction

Other Consultants
Davis Langdon & Seah
Cornell Wagner
LPA
Belt Collins

BEST TALL BUILDING

NOMINEE:
UNITY
LIVERPOOL
UNITED KINGDOM

page 82

WINNER

WINNER:
51 LIME STREET
LONDON
UNITED KINGDOM

page 64

NOMINEE:
SYV TOWER
MADRID
SPAIN

page 78

NOMINEE:
TORRE ESPACIO
MADRID
SPAIN

page 80

AWARDS 2008

EUROPE

NOMINEE:
HET STRIJKIJZER
THE HAGUE
THE NETHERLANDS

page 72

NOMINEE:
MADOU PLAZA
BRUSSELS
BELGIUM

page 76

NOMINEE:
COVENT GARDEN
BRUSSELS
BELGIUM

page 70

NOMINEE:
KANYON COMPLEX
ISTANBUL
TURKEY

page 74

2008 WINNER: BEST TALL BUILDING EUROPE

51 LIME STREET
LONDON, UNITED KINGDOM

51 Lime Street acknowledges that the way in which an office building responds to the context and spirit of the city in which it stands is as fundamental to its success as the acknowledged benefits of natural ventilation, light, open space and a view. As a result, the architects continue their built explorations into new strategies for flexible, column-free office space, but have also created the idea of the "urban room", where genuine connections to the public realm are established, and the way in which the building "touches" the ground is paramount.

51 Lime Street, also called the Willis Building (after its primary tenant), sits in the heart of the City of London. It lies to the east of Richard Rogers' 1986 Lloyds Building and responds to this unique location with an elegant concave form. The project is significant in both urban and environmental terms—51 Lime Street is among a number of projects in the City of London that have struck a delicate balance between commercial requirements, the need for flexibility, and respect for the area's world-famous architectural heritage. The original building for Willis Faber Dumas by Foster + Partners in 1976 was a seminal project for the practice, an open office building characterized by its sense of community—this spirit has been kept alive in the new UK headquarters in London, with expansive roof terraces that offer broad views over the city (*see Looking down at the rooftop terrace image on page 67*).

The development comprises two separate buildings which step down to a new public plaza (*see Ground floor plan on page 69*). The 10-story building at 1 Fenchurch Street responds to the smaller scale of Billiter Street and Fenchurch Avenue, while the 29-story Willis headquarters building rises to the west of the site. The smaller building's concave façade encircles the plaza and its curved corners maintain important view corridors, and also restore a pedestrian route through the site, reinforcing the medieval street pattern. The landscaping also features sculpture reclaimed from the previous building, linear benches and a living wall to replace the existing 'party wall' between the Willis Building and its neighbors, improving the view from the building and encouraging biodiversity (*see Ground level plaza image on page 66*). With a fringe of cafes, restaurants, shops and bars at the tower's base, 51 Lime Street extends

AWARDS JURY / EDITOR STATEMENT

"This is a project that integrates well with the urban context, both historic and modern. I appreciate its sensitivity to the environment through the building's 'serrated' exterior that simultaneously shades from the sun while letting views out."

-Tim Johnson, NBBJ

View from the north; image © Nigel Young/Foster + Partners

PROJECT DETAILS

Completion Date
April 2008

Height
125 meters/409 feet

Total Area
50,107 square meters/
539,347 square feet

Use
Office

Owner / Developer
The British Land Co Plc/
Stanhope Plc/ St Martins
Property Group

Architect
Foster + Partners

Structural Engineer
Ramboll Whitbybird

MEP Engineer
Roger Preston and Partners

Construction Manager
Mace Ltd.

Other Consultants
Davis Langdon & Seah
Mott Green Wall
Gordon Ingram Associates
Sandy Brown Associates
David Bonnett Associates
Charles Funke Associates
Equation Lighting Design
Emmer Pfenninger

51 LIME STREET
LONDON, UNITED KINGDOM

Ground level plaza; image © Nigel Young/Foster + Partners

the vibrancy of the nearby Leadenhall Market, a particularly lively shopping area in the City with a strong architectural character.

As towers grow ever taller, the strategies to achieve stability are increasingly central to the design approach. On plan, the Willis headquarters has been developed as a series of overlapping curved shells, while its section is arranged in three steps. The roof terraces overlooking the plaza on the lower two steps are directly accessible from the office spaces. Both buildings have a central core to provide open floor plates and maximum flexibility in use, so they are able to accommodate a number of different configurations for one or more tenants.

The entire development is visually unified by its highly reflective façade. The pressed form of the panels and their mica finish give them depth and texture. A strong language is established through the interplay of solid and glazed panels arranged in a saw-tooth pattern. The fins also increase insulation while reducing glare and solar gain—just one of the strategies that have contributed to the building's BREEAM "excellent" rating. Together with highly efficient services

The "saw toothed" facade reduces solar gain and glare; image © Nigel Young/Foster + Partners

AWARDS JURY / EDITOR STATEMENT

"The common design connections with Foster's original Willis Faber Dumas building 30 years previously are striking, specifically the building envelope which seems to both consume and reflect the urban context. Both projects book-end an amazing palette of work from Foster."

-Antony Wood, Editor

Looking down at the rooftop terrace; image © Nigel Young/Foster + Partners

51 LIME STREET
LONDON, UNITED KINGDOM

Main Elevator lobby; image © Nigel Young/Foster + Partners

Glass curtain wall on west façade; image © Nigel Young/Foster + Partners

equipment and systems and extensive bicycle parking, the building's progressive environmental strategy surpasses statutory carbon reduction targets by more than 20%.

Awards Jury Statement:

Europe is not known for very tall structures. This is largely due to significant historic fabrics developed over typically many centuries and the protection and respect needed for these areas. What the jury found most interesting in 51 Lime Street is that it is a project that integrates well with the urban context, both historic and modern. The project establishes a great dialogue with the existing fabric of historic masonry buildings and more modern glass buildings. The building accomplishes this though its dramatic form, its materiality, and by the articulation of the massing. The jury appreciated the design team's sensitivity to the environment. The building's "serrated" exterior simultaneously shades the façade from the sun and allows views out to the city. In addition, the formation of public gathering spaces and resulting activation of the ground plane further reinforces the CTBUH mission to see tall buildings integrated into the urban environment.

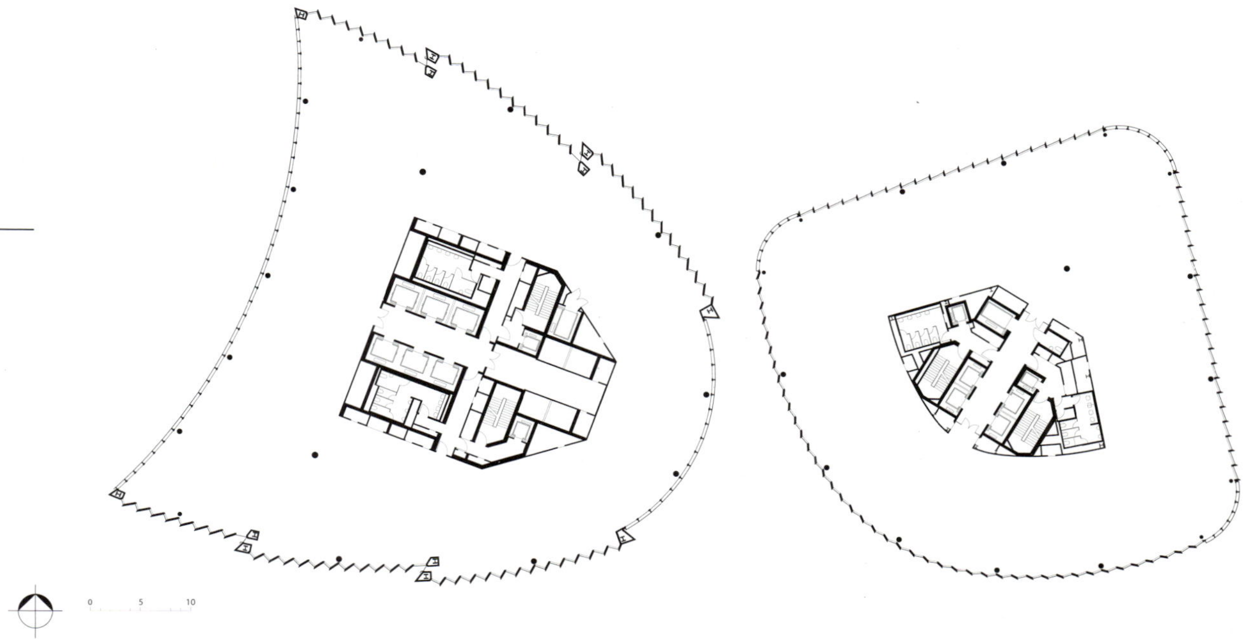

Third floor plan showing typical open office floor plates

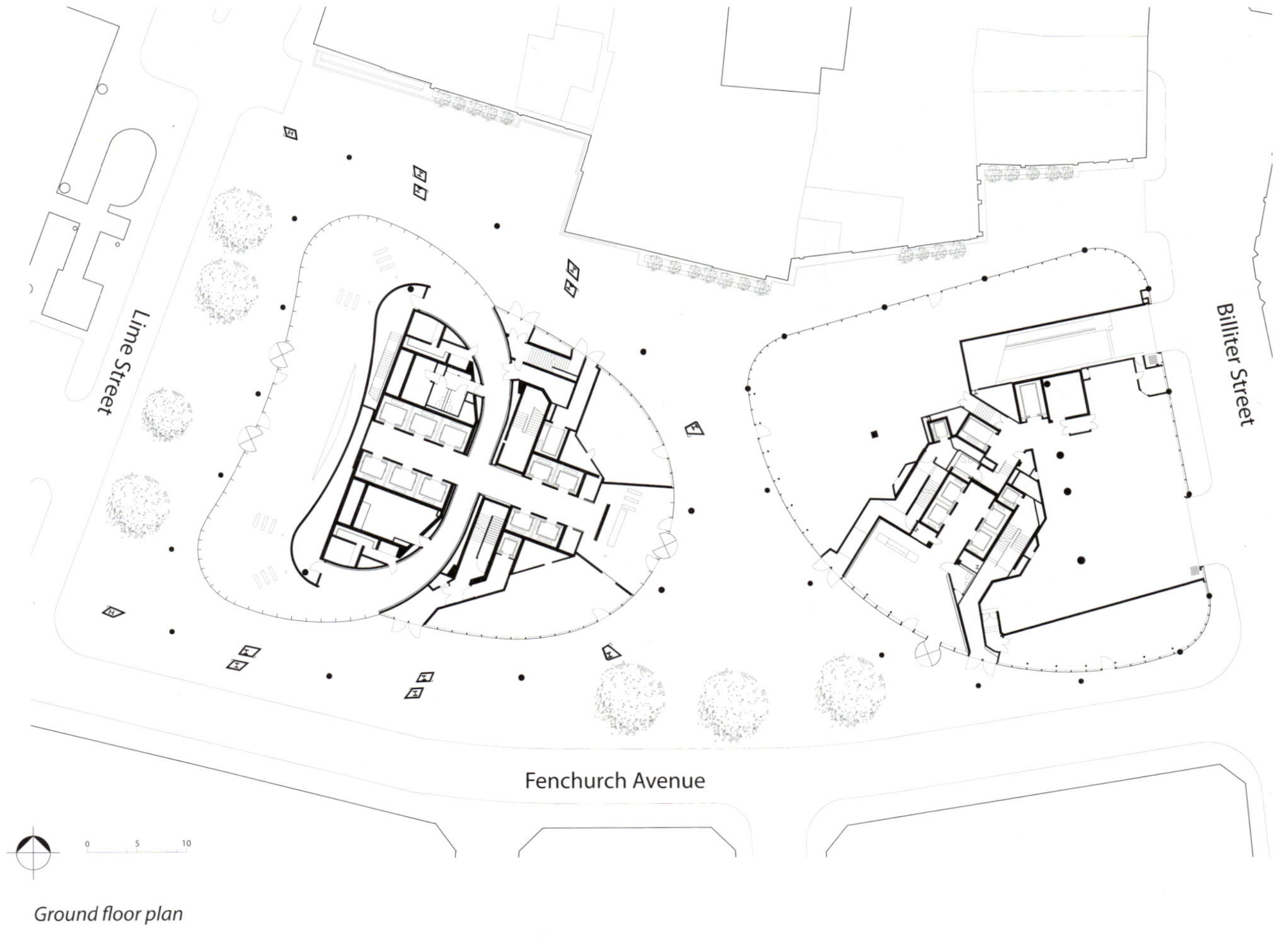

Ground floor plan

2008 NOMINEE: BEST TALL BUILDING | EUROPE

COVENT GARDEN
BRUSSELS, BELGIUM

Night view from the south

Covent Garden is comprised of a high-rise and low-rise building, meant to become an urban signal announcing the North area, the only purpose-built high-rise business district of Brussels. From an urban point of view, the project blends in as much with the historic low-rise buildings it faces on one side, as it copes with the tall buildings it faces on two other sides. At ground level a pedestrian link is created between an existing park and a major public square via a landscaped garden-atrium.

The project implements many sustainable features, including operable window-boxes that allow for natural ventilation yet do not allow rainwater into the building. A radiant air-conditioning system is used, in which less energy is used to achieve the same comfort level as a traditional air-conditioning system. The radiant air-conditioning system or "cold ceiling" requires negligible maintenance and a very low overall cost. Fresh air enters the offices through ventilation grills in the ceilings and is extracted near the lighting after being drawn across the offices.

Lastly, Covent Garden is equipped with an installation which in its specific application

"With water becoming an ever-more endangered resource, this project presents an innovative and responsible way to handle its own water, naturally treating it on site."

-Antony Wood, Editor

AWARDS JURY / EDITOR STATEMENT

is a first in Belgium: the garden of the atrium is not only a major meeting space within the complex, it is also an "Eco-Machine" meant to organize wastewater recovery.

The Eco-Machine is a wastewater treatment process that treats both grey and black water, using advanced biological and bacteriological purification techniques. The system proceeds initially by sedimentation in a septic tank and biological purification with nitrification: the bacteria are held in free suspension in the water and the biomass is filtered through membranes. In a second stage, the water transits via the covered garden into vats with helophytes, which play a fine purification role.

The objective is to treat this water so that it can be recycled into the building's own consumption cycle. Water is recovered at the end of the process and is stored in a pond. Although not drinkable, it is re-injected into the building for sanitary uses (washrooms), for building maintenance and for watering the plantings.

Atrium Garden where the plantings contribute to the wastewater treatment process

View looking up from the 21st floor open air patio

PROJECT DETAILS

Completion Date
April 2008

Height
99 meters/325 feet

Total Area
74,000 square meters/
796,529 square feet

Use
Office, Retail

Owner / Developer
Immobiliere Du Royal Rogier/ Covent Garden Development
Buelens Real Estate

Architect
Art & Build Architect sa/ Montois Partners Architects sa

Structural Engineer
B.E. Greisch/ Van Wetter

MEP Engineer
AM Jcd. Std.

Contractor
Van Roey/ DEMOCO/ Vooruitzicht/ Belgometal/ Mitsubishi/ Geo Milieu/ Van Den Rijse/ Close/ ITB

Other Consultants
Joost Valgaeren
Abssis
Atelier 50
Venac
Seco
Axioma

2008 NOMINEE: BEST TALL BUILDING | EUROPE

HET STRIJKIJZER

THE HAGUE, THE NETHERLANDS

Night view from the north; image © Peter de Ruig

Het Strijkijzer is a fully pre-cast residential tower with an L-shaped plan. Sited on a triangular site of only 37 meters by 34 meters (121 by 111 feet) at one of The Hague's busiest intersections, the building takes cues from The Flatiron Building in New York City (1902) in its form. The building contains 351 units: 51 luxury apartments and 300 studios intended to provide an affordable housing option to young adults and students. There are also many public features on the first three levels, and a Grand Cafe surrounded by an outdoor gallery on the top floor with separate elevator access.

With the small site completely occupied by the building, there was no room for a construction staging area, and as such innovative construction methods were developed. Two different companies prefabricated the concrete floor and wall pieces. The delivery trucks would have to meet on the highway and ensure they were driving in the correct order so that when they arrived at the site, the parts could be taken from the trucks and installed on the building immediately. This led to an incredible construction speed of two floors a week.

Interior corridor; image © Luuk Kramer

Exterior façade; image © Peter de Ruig

AWARDS JURY / EDITOR STATEMENT

"The similarities with New York's Flatiron Building are marked, now updated to inhabit 21st century Europe."

-Antony Wood, Editor

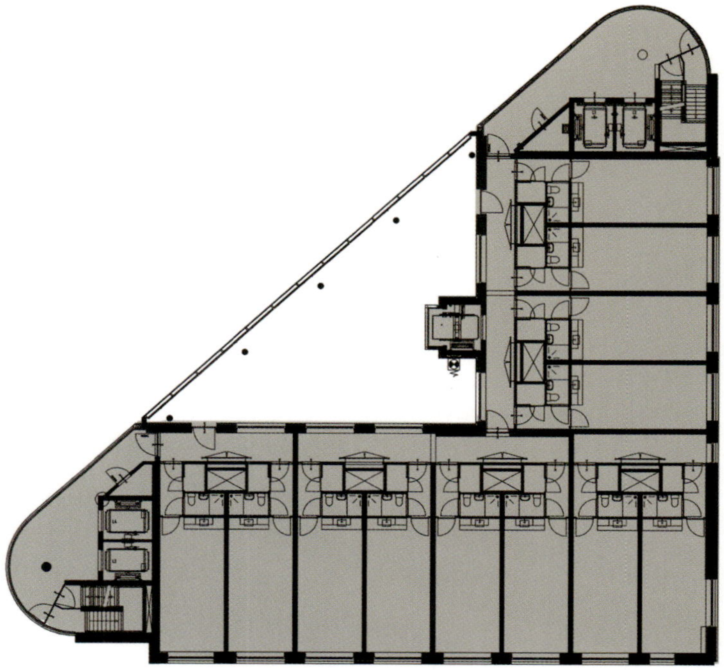

Typical floor plan at student housing levels

PROJECT DETAILS

Completion Date
October 5, 2007

Height
132 meters/433 feet

Total Area
30,000 square meters/
322,917 square feet

Use
Residential, Retail

Owner / Developer
Vestia

Architect
AAARCHITECTEN b.v.
A.P. Bontenbal

Structural Engineer
Corsmit Consulting Engineers

MEP Engineer
Deerns Consulting Engineers

Contractor
Boele & Van Esteren

Other Consultants
Aveco De Bondt
Bijlbouw
Ketlitz
Oosthoek/ Kemper
Oostingh Staalbouw
Twijnstra Gudde

2008 NOMINEE: BEST TALL BUILDING | EUROPE

KANYON COMPLEX
ISTANBUL, TURKEY

View of the main office tower

The Kanyon Complex is a city-within-a-city, a diverse mix of residential and office, retail stores, restaurants, and entertainment components designed to provide a stage for "urban theater" and promote interaction among its residents and visitors. This was accomplished by organizing the project as a series of districts—Entertainment Sphere, Office Plaza, Garden Court and Performance Plaza—connected by an interior street, or "canyon", that traverses the site and includes a variety of courtyards and terraces.

The functional relationships of the uses and their relationship to the city establish the basic planning armature for the project. The 25-story office building is located along a major boulevard to give it a primary presence within the city. The four levels of retail extend the full length of the site and create a podium for residences above. The retail has multiple entrances at different levels that knit the project to the city at various street locations and to the below-grade subway station.

The 179 apartment residences above the retail are formed in terraces; their location and massing give them both privacy and dramatic views into the project below. Kanyon's design is conceived as a composition of bold, memorable architectural forms, complemented by unique building materials. For example, the canyon flows across the project, carving out soft curves through the building's forms, featuring a stratification of light, earth-tone colored stone with accent bands of painted copper metal. The gentle curve of the office tower façade reinforces the graceful flow of the canyon and is sheathed in a glass curtain wall.

AWARDS JURY / EDITOR STATEMENT

"The urban scale of this development, and the relationship between tower and landscaping (both physically and materially) is interesting. 'Kanyon' is an apt name for it."

-Antony Wood, Editor

View of the "canyon" that traverses the site

Outdoor lighted dome within the retail area of the complex

PROJECT DETAILS

Completion Date
June 2006

Height
118 meters/387 feet

Total Area
250,000 square meters/
2,690,980 square feet

Use
Office, Residential, Retail

Owner / Developer
Eczacıbaşı Group and İŞGYO of Turkey

Architect
The Jerde Partnership

Associate Architect
Tabanlıoğlu Architects

Structural Engineer
Arup

MEP Engineer
Arup

Contractor
Tepe Construction

2008 NOMINEE: BEST TALL BUILDING | EUROPE

MADOU PLAZA
BRUSSELS, BELGIUM

View from the southwest

Madou Plaza, originally completed in 1965, is the tallest building in the Brussels CBD. Its renovation is probably one of the most ambitious of its kind ever undertaken in the city. Completely rethought in terms of its function, circulation routes and relationship to its surroundings, the 34-story building has been entirely rebuilt around its original structure.

Some 8,000 square meters (86,111 square feet) of new construction at the base have created a podium for the existing tower. This 13-level podium is organized around an atrium, the tallest in Brussels. Madou Plaza now includes a conference center which is also accessible to the general public and equipped with an auditorium and business services facilities.

The "glazed promontory" located in the façade underlines the design of the new entrance to the tower; the micro-climate of the plaza is improved by both the feature and the design of the curved façades.

A semi-reflective double-skin type façade and cold ceilings provide the occupants with the highest standard of comfort. The entire ceiling surface acts as a cooling radiator thanks to the circuit of chilled water in tiny tubes circulating behind the perforated metal panels of the ceiling.

At ground level, a large area along a retail street has been allocated to shops. A new street flanked by housing (to be built soon) has been created at the rear of the tower while an esplanade has been laid out at the front. Madou Plaza now constitutes a genuine plaza, which, for the past 20 years, had existed in name only.

AWARDS JURY / EDITOR STATEMENT

"The images of before and after tell the whole tale. In today's climate-challenged world of increasingly precious embodied energies, this project points the way for many existing tall buildings in the future."

-Antony Wood, Editor

Tower prior to renovation

View down in 13-story atrium

PROJECT DETAILS

Completion Date
1965 (original)
2007 (renovation)

Height
120 meters/394 feet

Total Area
40,000 square meters/
430,556 square feet

Use
Office, Retail, Auditorium

Owner / Developer
Commission of the European Union/ IVG Real Estate Belgium

Architect
ASSAR Architects

Associate Architect
Archi 2000

Structural Engineer
Waterman TCA

MEP Engineer
Tractebel Development Engineering

Contractor
Herpain Entreprise/
Strabag

Other Consultants
JNC International

2008 **NOMINEE:**

BEST TALL BUILDING | EUROPE

SYV TOWER
MADRID, SPAIN

View from the south; image © Rafael Vargas

SyV Tower is part of a set of four towers located in the north of Madrid on the Paseo de la Castellana, a major boulevard in the city (*Torre Espacio on page 80 is also part of this complex*). With Madrid's lack of tradition in high rise construction and the site of these towers being on the highest topographic area in the city, they create a dramatic new impression on the skyline. SyV Tower stands out among them with its clarity of volume. The building is fragmented by fissures into three pieces, which increase the volume's sense of verticality. These fissures introduce light into the building's interior creating the illusion of grouping single vertical pieces. The shape of the floor plan emerges from the study of both the minimum surface resistance against the wind, and the optimal relationship between usable area and façade length.

The structural scheme consists of two concentric rings of pillars of variable cross section beginning with a circular cross section, changing to a rectangular cross section in the hotel levels as they integrate more easily into the enclosure, and then returning to a circular cross section again in the open floor plans of the offices. A Y-shaped central core of reinforced concrete houses the elevators and service shafts in the center.

The tower is wrapped entirely by a double-skin façade. It was considered fundamental, given the climate conditions of the city of Madrid, to protect the building from excessive solar radiation. In the same way of traditional Spanish architecture, the façade solves the problem by creating shade over the building. The floor plates extend beyond the inner glass wall to create overhangs, and the outer glass wall is open and permeable, arranged in slightly overlapping shingles. The variation of angles between these "glass shingles" generates small local distortions creating a light layer of turbulent air that helps wind slide through with minimal friction.

AWARDS JURY / EDITOR STATEMENT

"The outer permeable skin, with its overlapping vertical glass 'shingles' is a remarkable design feature, elevating the curtain wall way beyond the banal."

-Antony Wood, Editor

View of space between the two layers of the double façade; image © Roland Halbe

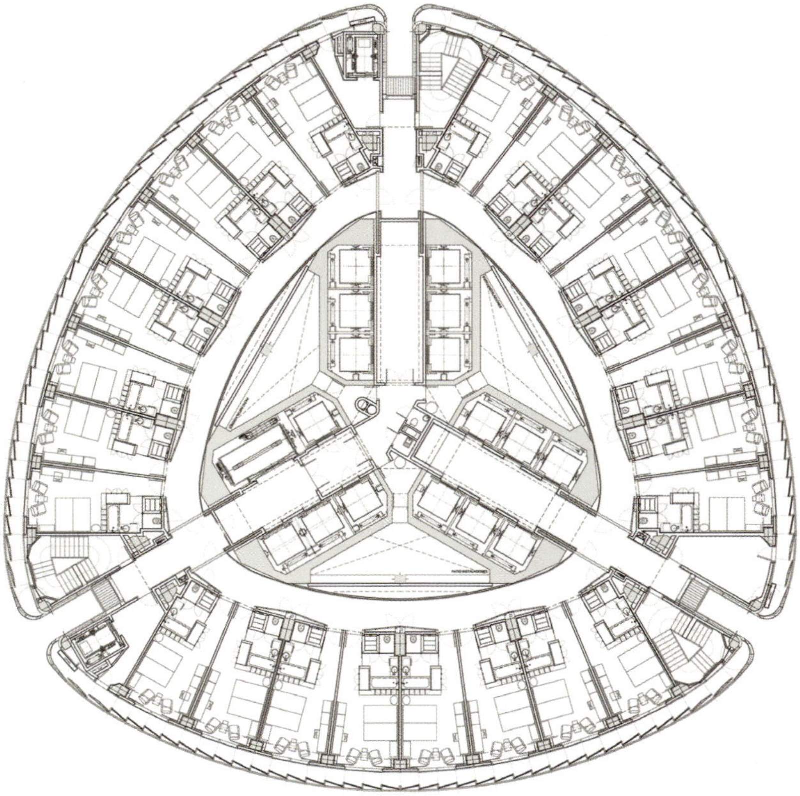

Typical hotel floor plan

PROJECT DETAILS

Completion Date
October 2008

Height
236 meters/774 feet

Total Area
110,000 square meters/
1,184,030 square feet

Use
Office, Hotel

Owner / Developer
TESTA Inmuebles en Renta S.A.

Architect
Rubio & Álvarez-Sala estudio de arquitectura

Structural Engineer
MC2 Estudio de Ingenieria

MEP Engineer
Aquilera Ingenieros / Urculo Ingenieros

Contractor
SACYR

Other Consultants
Estrumaher
Schlaich Bergermann und Partner
Bureau Veritas
INTEMAC

2008 NOMINEE:
BEST TALL BUILDING | EUROPE

TORRE ESPACIO
MADRID, SPAIN

Night view from the east

Torre Espacio is designed to energize the space in which it stands and animate the skyline of Madrid. The design of Torre Espacio evolves from a square plan at its base to a gently curving lozenge, shaped by the intersection of two quarter circles, at its crown. Its intersecting curved surfaces are composed of many separate panels that must be individually fabricated and assembled. Hence, there was a need to discover and adopt a geometric order that can rationalize and facilitate the desired evolution from square to quarter circle. The architects have found this order by subdividing the building section into 44 horizontal layers and then subdividing each quarter circle of the plan into an equal number of radial segments. The intersections of successive layers in section with successive segments in plan define points along a vertical curve.

The mathematical formula describing this curve is based on the cosine, and while it simplifies fabrication and assembly of the curtain wall, its rate of curvature is not constant, but rather changes gradually over its entire length. This gives the resulting tower form its distinctive and visually appealing sense of energy.

"Torre Espacio demonstrates the possibilities now open to the designer yearning to move away from orthogonal form and fabrication."

-Antony Wood, Editor

AWARDS JURY / EDITOR STATEMENT

The façade is an internally ventilated climate wall design. It achieves an average of 1.3W/m2C thermal rating. It integrates façade technology with the building HVAC systems. This results in a reduction of energy consumption, improved occupant comfort and acoustical properties, and a high transparency ratio which allows the façade to be executed in floor-to-ceiling glass. Better utilization of natural light reduces artificial lighting, leading to direct energy savings. Reduced cooling loads from less artificial lighting also leads to indirect energy savings.

The design also utilizes cool ceilings, with a radiant grid of chilled water looped above the perforated metal ceiling panels which increases the comfort in the office spaces by cooling without drafts.

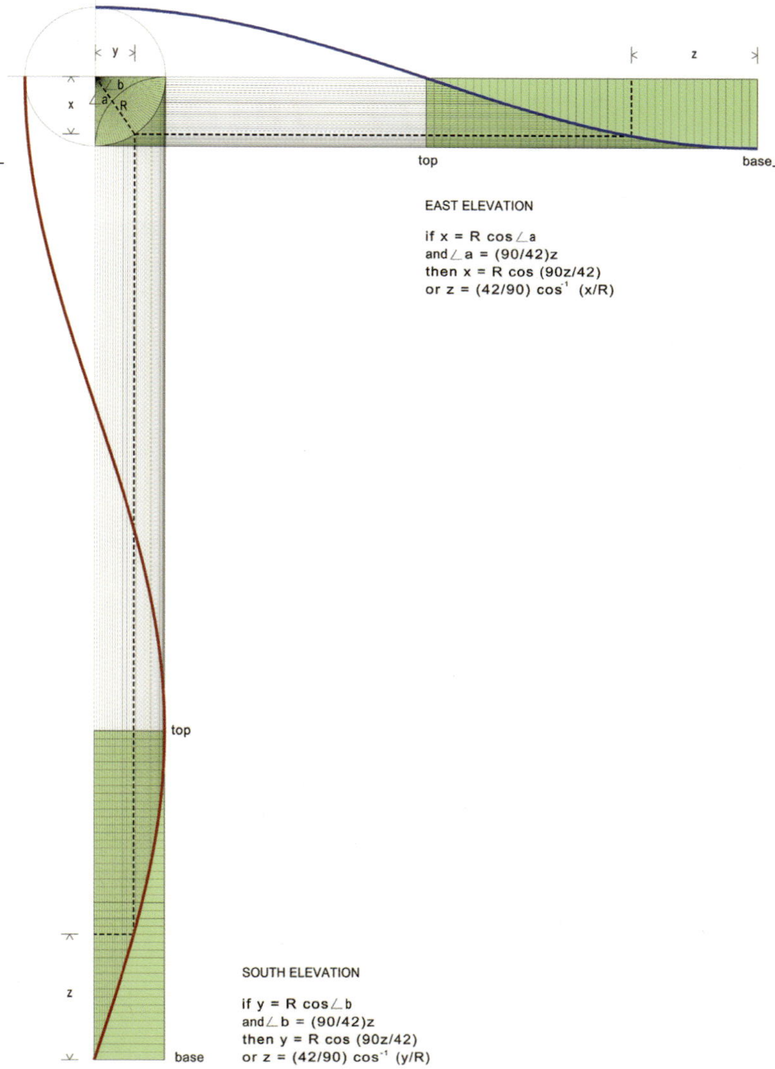

Diagram showing the plan and east and south elevations (in green) in relation to the cosine curves used to generate their form

PROJECT DETAILS

Completion Date
April 2008

Height
224 meters/735 feet

Total Area
7,500 square meters/
80,729 square feet

Use
Office, Retail

Owner / Developer
Torre Espacio Castellana, S.A.

Architect
Pei Cobb Freed & Partners Architects LLP

Associate Architect
Reid Fenwick Associados

Structural Engineer
MC2 Estudio de Ingenieria

MEP Engineer
R. Urculo Ings Consultores Associados, S.A.

Contractor
OHL Madrid

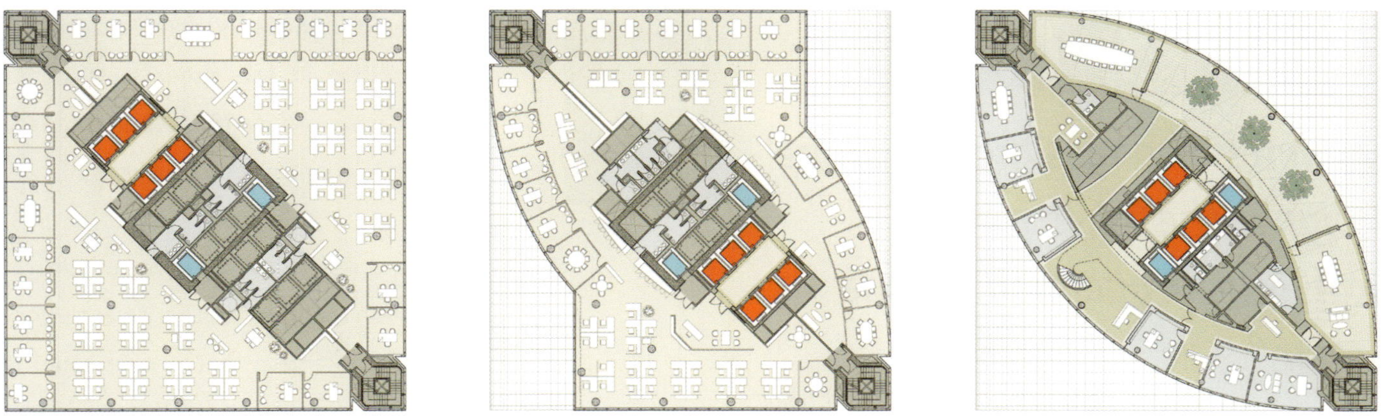

From left to right: 6th, 26th and 51st floor plans showing the progression from square to ellipse form

2008 NOMINEE: BEST TALL BUILDING | EUROPE

UNITY
LIVERPOOL, UNITED KINGDOM

View from the north

Unity is a mixed use scheme comprised of two towers, situated to the north of the iconic Liver Building. The two towers of the Unity building serve as powerful complements to the waterfront silhouette of the famous Merseyside buildings and the scheme has now become a symbol of the city's renaissance as 2008 European City of Culture.

The taller 27-story residential tower is crowned by a series of step-backs over which stands the stainless steel penthouse pod. It has 161 apartments mostly in a two-level maisonette crossover formation, the adjacent tower houses 16 stories of office space. A key urban intention of the design was to reinstate the pattern of street and city block which had been locally eroded. At street level the scheme is built to the property boundaries like any traditional urban block. Shops and commercial lettings are located in the street level podium which wraps around three edges of the site enclosing a landscaped courtyard.

In response to the massive middle class migration into the city, Unity was designed with the intention that people could live and work in the same neighborhood which

> ***"This building completely changes the landscape of tall building design in the UK, in its bold embrace of atypical form, proportion, façade articulation and color."***
>
> *-Antony Wood, Editor*

AWARDS JURY / EDITOR STATEMENT

allows the area to remain active 24 hours a day, 7 days a week.

Externally, elevations to the residential tower are made up of a pattern of heavily perforated cladding panels in elongated T configurations. The slightly more conventional office tower skin is based on horizontal strip windows and spandrel panels overlaid by a regular hit and miss pattern of projecting vertical floor-height glossy black panels. This patterning explores ideas of scale and urban camouflage, providing texture and visual interest to these large-scale urban interventions, and relate to the days of the Great War when 'dazzle ships' were painted on the Mersey River.

Residential tower and penthouse "pod"

Viewed in the Liverpool skyline from the Mersey River

PROJECT DETAILS

Completion Date
January 2007

Height
86 meters/282 feet

Total Area
17,364 square meters/
186,905 square feet

Use
Residential, Office, Retail

Owner / Developer
Rumford Investments

Architect
Allford Hall Monaghan
Morris LLP Architects

Structural Engineer
Faber Maunsell

MEP Engineer
Hoare Lea

Contractor
Laing O'Rourke

Other Consultants
Goyne Adams
John Shreeves & Partners
Sol Acoustics

BEST TALL BUILDING

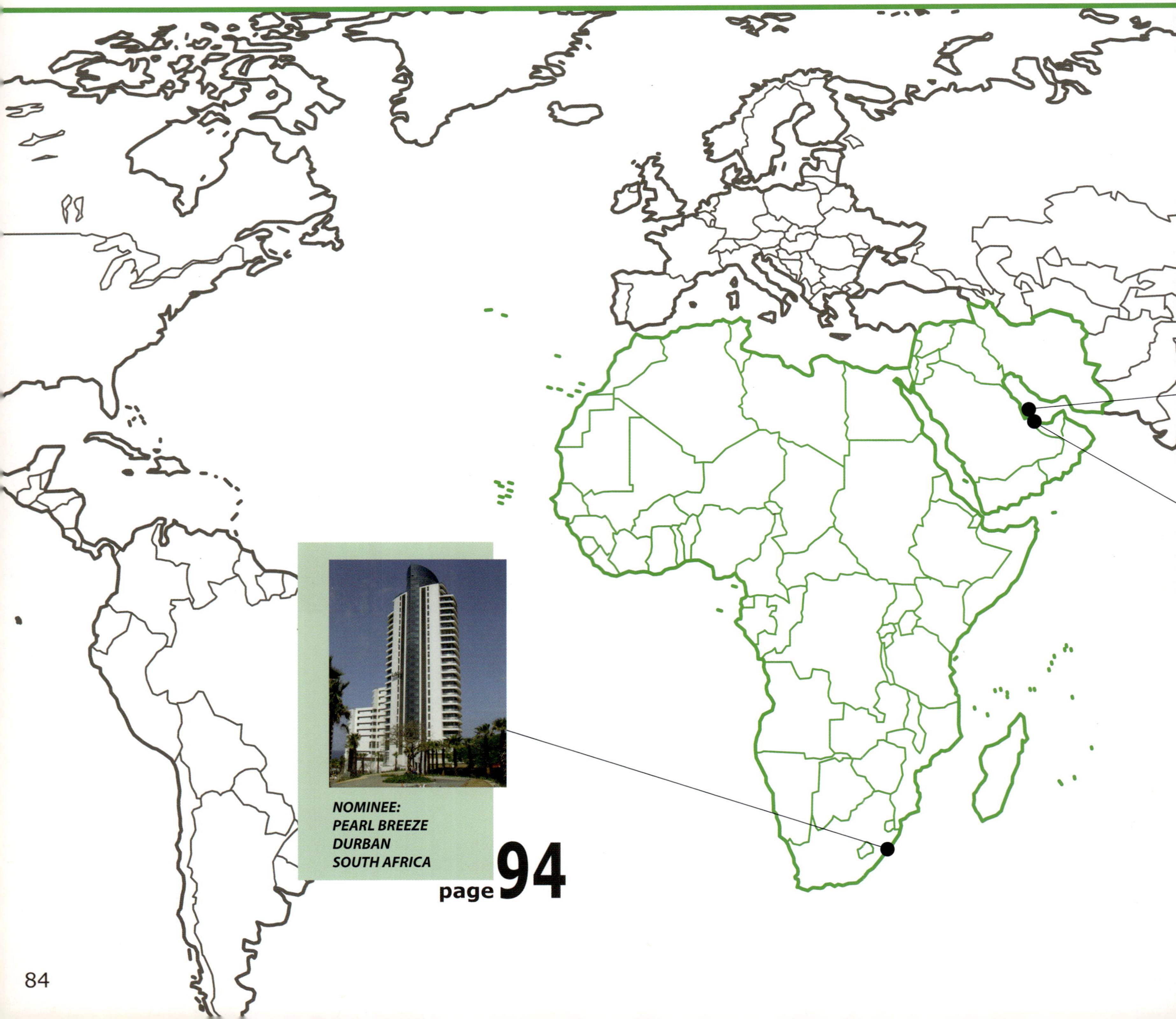

NOMINEE:
PEARL BREEZE
DURBAN
SOUTH AFRICA

page 94

AWARDS 2008

MIDDLE EAST & AFRICA

WINNER

WINNER:
BAHRAIN WORLD TRADE CENTER
MANAMA, BAHRAIN

page 86

NOMINEE:
ASPIRE TOWER
DOHA
QATAR

page 92

WINNER:

2008 BEST TALL BUILDING MIDDLE EAST & AFRICA

BAHRAIN WORLD TRADE CENTER

MANAMA, KINGDOM OF BAHRAIN

The Bahrain World Trade Center is the world's first building to integrate large-scale wind turbines; and together with numerous energy reducing and recovery systems, this development shows an unequivocal commitment to raising global awareness for sustainable design. This building is pioneering a new direction for designers and owners acting as a technological precedent. The BWTC has shown that commercial developments can be created with a strong environmental agenda and addresses the needs of our future generations. The BWTC encapsulates the essence of a sustainable philosophy engaging all of the social, economic and environmental impacts of the project. As well as making significant strides in environmentally balanced architecture, the building is now considered a source of national pride for Bahrain residents, and is attributed with generating economic prosperity within the capital of Manama.

The BWTC forms the focal point of a master plan to rejuvenate the 30-year-old existing hotel and shopping mall on the site. The planning of the site became constrained by the existing buildings and the road network around the site. By extending the main axis of the existing shopping mall towards the sea and creating a secondary axis from the Hotel, "Retail Streets" were established. The twin towers' natural location was therefore positioned on the main axis, facing the Arabian Gulf and creating the entrance for the development.

The inspiration for the 42-story twin towers originated from regional "Wind Towers" and their ability to funnel wind, and the vast sails of the traditional Arabian Dhow as they harness the breeze in driving them forward. After careful Computational Fluid Dynamics (CFD) modeling and extensive wind tunnel testing, the towers' shape was literally carved out by the wind to create optimum airflow around the buildings. The elliptical plan forms act as aerofoils (*see plans on page 91*), funneling the onshore breeze between them, creating a negative pressure behind, thus accelerating wind velocity between the two towers. Vertically, the sculpting of the towers is also a function of airflow dynamics.

As they taper skywards, the aerofoil sections reduce (*see elevations on page 90*). This effect, combined with the increasing onshore wind velocity at higher altitudes

AWARDS JURY / EDITOR STATEMENT

> ***"The Bahrain World Trade Center, with its three massive turbines, gives a very strong visual nod toward sustainability."***
>
> *-James Forbes, Hyder Consulting Pty. Ltd.*

Looking up from the main entrance at the wind turbines

PROJECT DETAILS

Completion Date
April 2008

Height
240 meters/787 feet

Total Area
120,000 square meters/
1,291,669 square feet

Use
Commercial, Retail

Owner / Developer
Confidential

Architect
WS Atkins & Partners

Structural Engineer
WS Atkins & Partners

Contractor
Nass, Murray & Roberts (JV)

BAHRAIN WORLD TRADE CENTER

MANAMA, KINGDOM OF BAHRAIN

View of one of one of three 29-meter-diameter wind turbines

Central court of the retail podium

creates a near equal regime of wind speed on each of the three turbines, irrespective of height, allowing them to rotate at the same speed and generate the same energy levels. The three 29-meter (95-foot)-diameter, 11 ton wind turbines are supported on 31.5-meter (103-foot), 70 ton bridges between the towers. Each turbine generates 225kW. The buildings have been sculpted to funnel the uninterrupted on-shore breeze onto the turbines and create a perpendicular slip stream that corrects the wind direction to take advantage of 70% of Bahrain's wind energy.

The premium on this project for including the wind turbines was less than 3% of project value. Based on the energy savings and the increased value of the building having wind turbines, the payback period is extremely favorable. The initial energy yields during the design phase was approximately 15%, therefore 1300MWh per year; however, from early commissioning results the turbines are estimated to generate substantially more energy due to the reduced occupancy profile of the building and the wider operational period of the turbines.

Interior view of a podium skylight

AWARDS JURY / EDITOR STATEMENT

"Perhaps the bravest high-rise adventure in a decade since Foster's Commerzbank, this project stands as testimony to the challenges facing us , the options opening up to us, and the growing number of courageous clients and consultants being revealed to us."

-Antony Wood, Editor

Towers viewed from the north on the waterfront

BAHRAIN WORLD TRADE CENTER

MANAMA, KINGDOM OF BAHRAIN

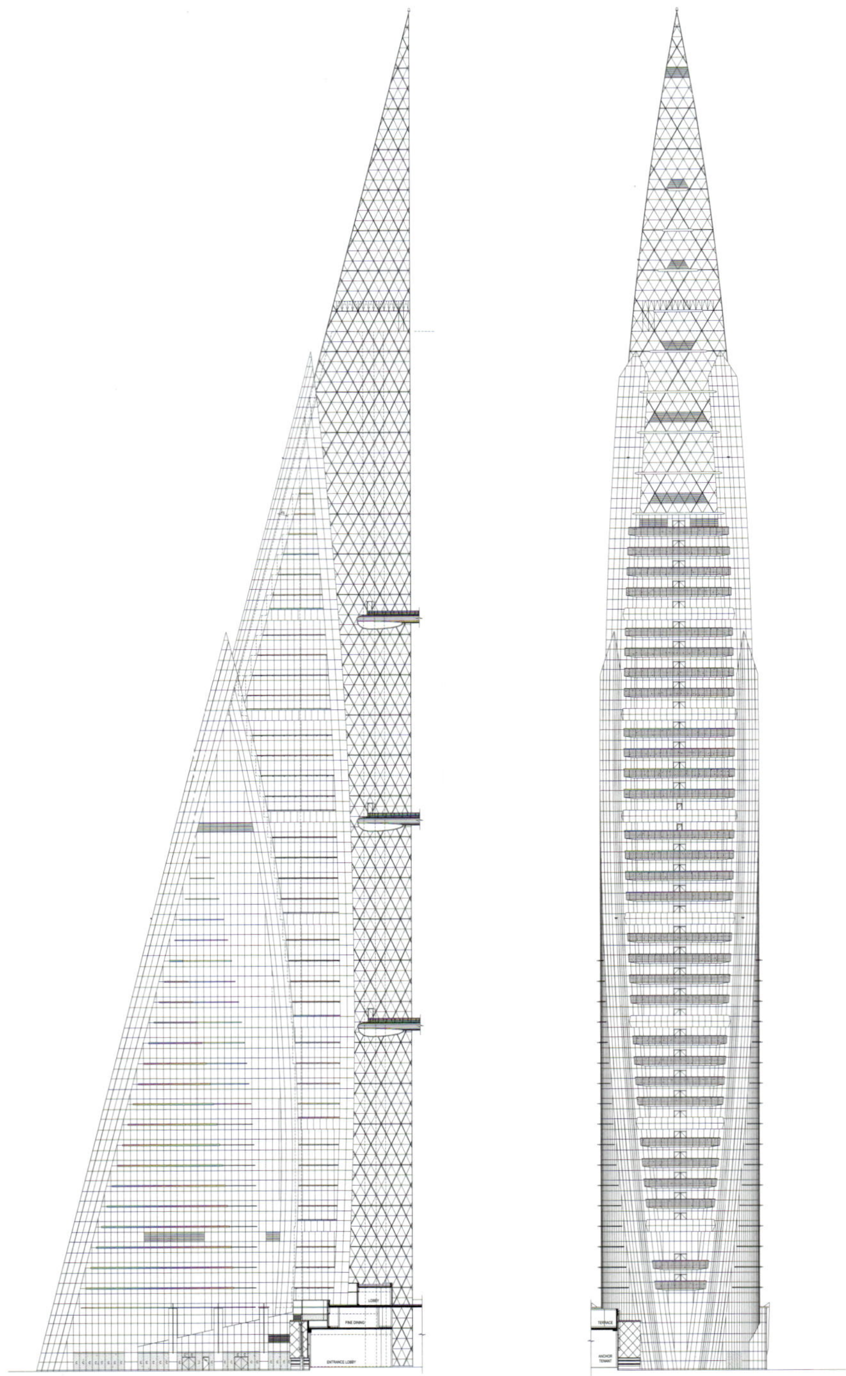

Tower elevations

Awards Jury Statement:

The Middle East is booming with high rise construction. We are all in awe of the great structures that are emerging from this region. At the same time, we must all take pause to assure ourselves that these cities are being properly planned, designed, and built and that they will contribute to the environment, not detract from it. The jury's deliberation of submissions from this region were looking for projects that simultaneously embodied strong bold building statements and at the same time integrated sustainable features into the design in this dramatic climate. The Bahrain World Trade Center was the project we felt embodied both. This project is a dramatic solution to a dynamic place in today's rapidly developing world. Its graceful form takes reference from its local geography and seafaring culture. In addition, the owner and design team made a valiant effort to integrate sustainable features into the building and express them as part of the architecture. The highly visible and dramatic wind turbines were seen as a strong iconic statement about the importance of alternative energy sources.

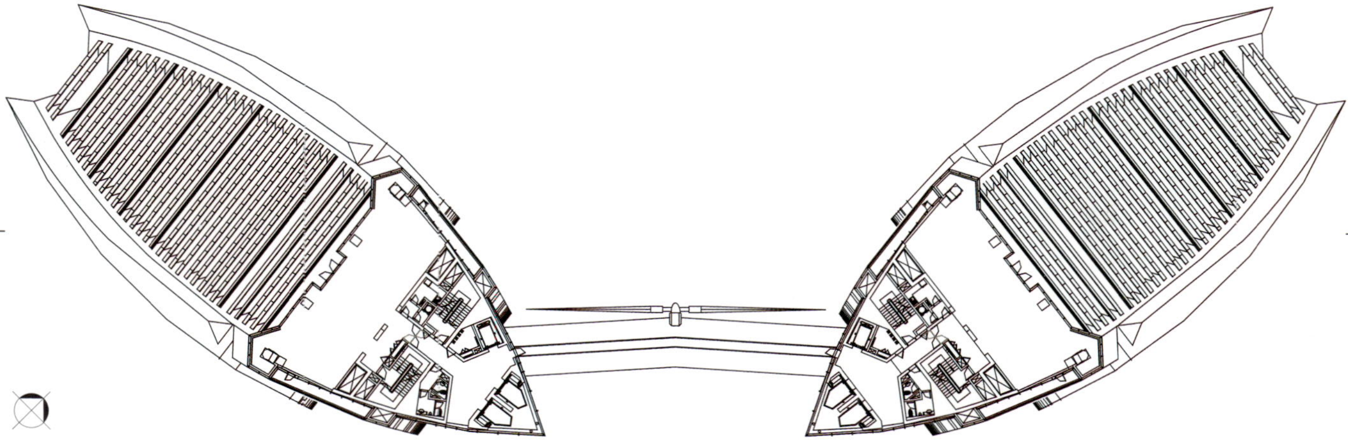

37th floor plan—upper turbine level

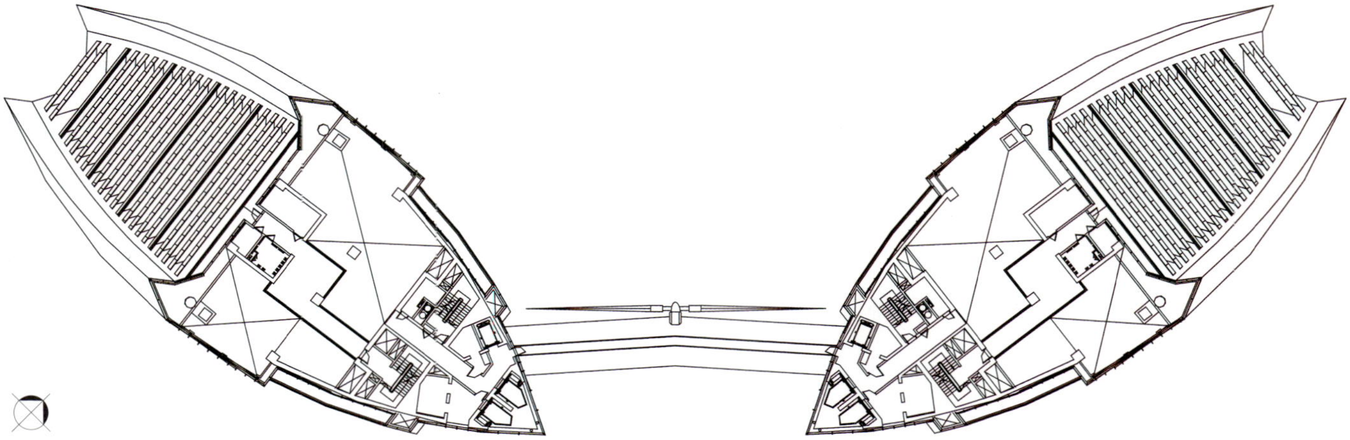

27th floor plan—middle turbine level

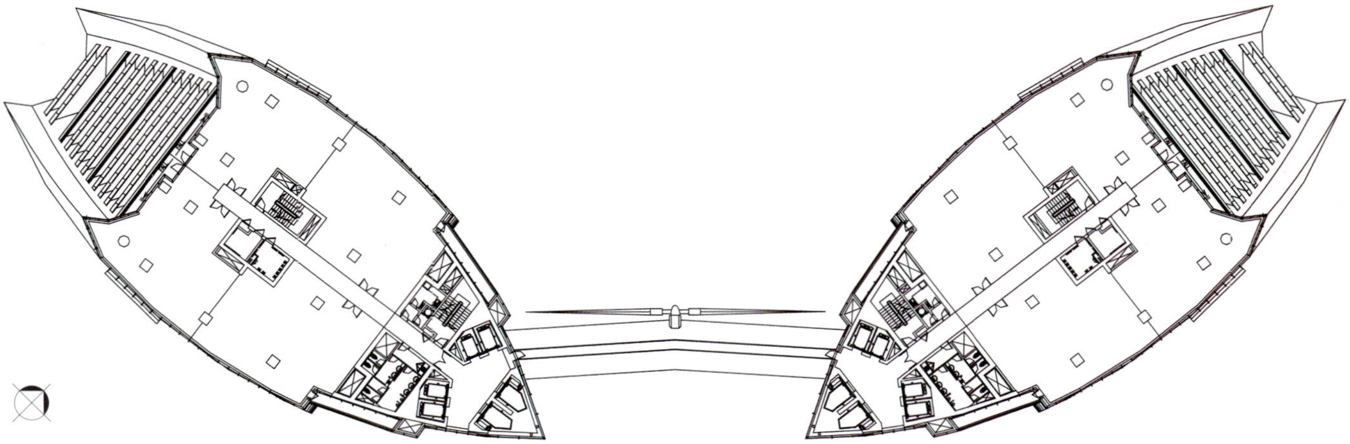

17th floor plan—lower turbine level

2008 NOMINEE: BEST TALL BUILDING | MIDDLE EAST & AFRICA

ASPIRE TOWER
DOHA, QATAR

Night view

The Aspire Tower was built to house the Symbolic Flame of the 2006 Asian Games. (The structure was completed in 2006 in time for the games, but the building was not completed and operational until 2007). The design symbolizes a hand grasping the torch that sits at the tower's top. The tower houses a five-star hotel, a sports museum, a health club with a cantilevered swimming pool, a revolving restaurant, and an observation deck.

The structural design features a concrete core enveloped in a steel mesh. The cone and lattice shell that house the torch rest atop this structure. The core supports the clusters of floors, which are all cantilevered up to 11.3 meters (37 feet) out from the core. The primary support beams extend radially from the core to steel columns on the outer edge of the structure.

The building features a swimming pool 80 meters (260 feet) above ground that is cantilevered from the core, and extends out 12 meters (39 feet) from the façade. Elliptical in plan, the pool is supported on a substantial steel truss structure some 4 meters (13 feet) deep. The plan geometry of the truss is straight from the core to the column locations, and then elliptical with approximately the same shape as the pool. The truss is connected to the core and supported on two columns that continue down through the levels below, which are supported on vertical trusses.

The façade consists of an energy-efficient glass skin, and helps efficiently maintain comfortable temperature levels in the desert climate.

AWARDS JURY / EDITOR STATEMENT

"The Aspire Tower pushes tall building design down an avenue it needs to tread, namely how to encourage non-traditional functions (in this case sports) in the high rise domain."

-Antony Wood, Editor

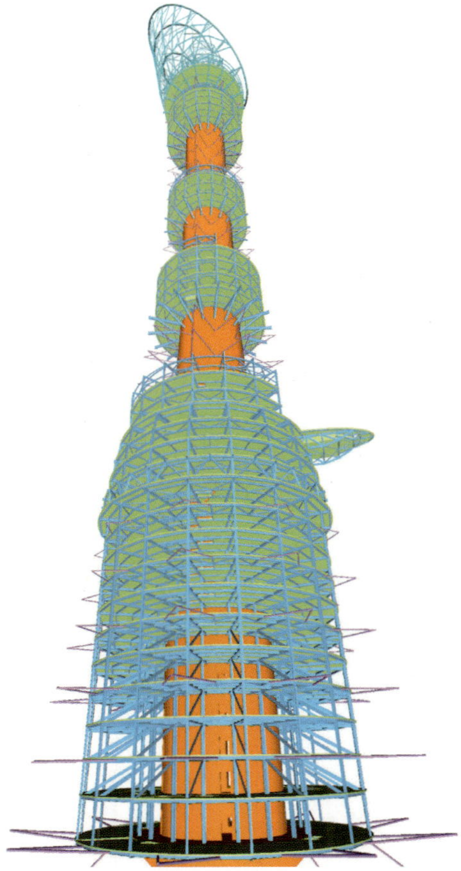

Structural diagram showing floors cantilevering from the concrete core (shown in orange)

Crown by day

PROJECT DETAILS

Completion Date
Spring 2007

Height
300 meters/984 feet

Total Area
39,948 square meters/
430,000 square feet

Use
Hotel, Retail

Owner / Developer
Doha City

Architect
Hadi Simaan architects/
AREP

Structural Engineer
Arup

MEP Engineer
Arup

Contractor
Midmac-Six Construct JV

Other Consultants
Ecart

2008 NOMINEE: BEST TALL BUILDING | MIDDLE EAST & AFRICA

PEARL BREEZE
DURBAN, SOUTH AFRICA

View of tower

Pearl Breeze is part of the residential landmark complex in Umhlanga, Durban, known as "The Pearls of Umhlanga." The resort town of Umhlanga, "The Place of Reeds," is situated 20 minutes from Durban along the Dolphin Coast, the playground of bottlenose dolphins. Complete with unspoiled golden beaches and lush sub-tropical greenery Umhlanga is fast becoming a popular holiday destination for international tourists.

Pearl Breeze is the first tower to be completed in the complex, and two more high-rise towers are planned. At 24 stories Pearl Breeze is the shortest tower for the complex, the others Pearl Dawn and Pearl Sky, will stand at 28 and 39 stories respectively. Once completed the Pearls of Umhlanga will house the tallest towers to be built in the region. Also part of the complex is an apartment building that was renovated to house luxury accommodation and low-rise "cabanas" built on the sea-facing boundary of the site. The complex will also feature a state of the art health spa and a retail boulevard.

A design feature of the complex is the central landscape area which is overlooked

"This project is hopefully a sign of Africa's ability to embrace the high-rise as a sustainable urban form."

-Antony Wood, Editor

AWARDS JURY / EDITOR STATEMENT

and surrounded by the three towers. This area contains rolling lawns with tropical vegetation and a large feature swimming pool. The landscaped area also serves as the roof to the basement parking below. To enhance the development there is no on-grade parking with all underground parking accessed near the main entrance gate to minimize the visible area of roadways.

The project will help bring Durban into an exciting new age of modern architecture and will promote development in the area. Property values in this area have increased by more than 100 percent since 2000, and the trend is expected to continue with the introduction of new buildings and complexes such as the Pearls. This complex provides state of the art, luxury residences and facilities and will definitely be a destination for those living in and visiting the area. The proximity of the residences to the health spa and to the retail boulevard reduces the need for vehicles for transportation, therefore helping to reduce emissions in the area.

View of Pearl Breeze with low rise "cabanas" in the foreground

Aerial view rendering of the completed complex from waterfront

PROJECT DETAILS

Completion Date
June 2008

Height
118 meters/388 feet

Total Area
42,000 square meters/
452,084 square feet

Use
Residential

Owner / Developer
Pearls of Umhlanga
Development (Pty) Ltd.

Architect
Cooper Architects
Seedat & Seedat Architects
Creative Kingdom

Structural Engineer
Arup

MEP Engineer
BFBA Consultants

Contractor
Group Five/ Pandev JV

Other Consultants
Incorporated
Walters & Simpson/ LDM JV
Keith Africa
Uys & White
SIP Project Managers

LIFETIME ACHIEVEMENT

LYNN S. BEEDLE AWARD

The award recognizes an individual who has made extraordinary contributions to the advancement of tall buildings and the urban environment during his or her professional career. These contributions and leadership are recognized by the professional community and have significant effects, which extend beyond the professional community, to enhance cities and the lives of their inhabitants. The individual's contributions may be well known or little known by the public and may take any form, such as completed projects, research, technology, methods, ideas, or industry leadership.

The candidate may be from any area of specialization, including, but not limited to: architecture, structure, building systems, construction, academia, planning, development or management and may or may not be a member of the Council (CTBUH).

The contributions of the award recipient should be generally consistent with the values and mission of the CTBUH and its founder, Dr. Lynn S. Beedle.

The award is not intended to be awarded posthumously, although it may be so awarded in some cases.

Photo of Lynn S. Beedle

AWARDS 2008

FAZLUR RAHMAN KHAN MEDAL

This award recognizes an individual for his/her demonstrated excellence in design and/or research that has made a significant contribution to a discipline(s) for the design of tall buildings and the built urban environment. The contribution may be demonstrated in the form of specific technical advances, innovations, design breakthroughs, building systems integration or innovative engineering systems that resulted in a practical design solution and completion of a project(s). The consideration may be based on a single project or creative achievement through multiple projects.

The candidate may be from any discipline in design or research and may or may not be a member of the Council (CTBUH).

The personal attributes of the award recipient should be of high professional integrity and the individual's contributions should generally be consistent with the mission of the Council (CTBUH).

Note: The Award is not intended only for structural engineers. In fact, this award is intended to emphasize the unique multi-disciplinary nature of the Council and set it apart from other professional organizations' awards for single disciplines.

CTBUH FELLOW

CTBUH Fellows are recognized for their contribution to the Council over an extended period of time, and in recognition of their work and sharing of their knowledge in the design and construction of tall buildings and the urban habitat.

Photo of Fazlur Rahman Khan

WINNER:

2008 LYNN S. BEEDLE ACHIEVEMENT AWARD

CESAR PELLI
PELLI CLARKE PELLI ARCHITECTS

Cesar Pelli was born in Argentina where he earned a Diploma in Architecture from the University of Tucumán. He first worked in the offices of Eero Saarinen serving as Project Designer for the TWA Terminal at JFK Airport in New York and Morse and Stiles Colleges at Yale University. After this apprenticeship, he was Director of Design at DMJM and, later, Partner for Design at Gruen Associates. During this time, he designed several award-winning projects, including the San Bernardino City Hall in San Bernardino, California; the Pacific Design Center in Los Angeles; and the United States Embassy in Tokyo.

His history of designing tall buildings dates to 1977, when he became Dean of the Yale School of Architecture and founded Cesar Pelli & Associates (now Pelli Clarke Pelli Architects). The firm's first project was an expansion of the Museum of Modern Art, an undertaking that included the addition of a 65-story residential tower—his tallest building up to that point (*see page 100*).

Over the years, he has designed some of the world's best known tall buildings. They include the Petronas Towers in Kuala Lumpur (*see right*)—the world's tallest buildings from 1998 to 2004—as well as Two International Finance Center in Hong Kong, One Canada Square at Canary Wharf, London (*see page 101*) and the tower headquarters of the Bank of America in Charlotte, North Carolina. Recent towers include the Bloomberg Tower in New York City, Torre de Cristal in Madrid, Grand Torre Costanera in Santiago and Transbay Transit Center and Tower in San Francisco (*see pages 102-103*).

His tall buildings, like all of his projects, do not belong to a single style. Mr. Pelli believes that buildings should be responsible citizens and that each design is shaped by its context. As he has explained, "A building must be both background and foreground. As foreground, it must have some exceptional qualities. But it must also try very hard to knit into the fabric of the city."

Since resigning as Dean of Yale in 1984, he has continued to write and lecture on architecture. His writing includes the 1999 book *Observations for Young Architects*, published by Monacelli Press.

The recipient of twelve Honorary Degrees,

Night view of Petronas Towers, Kuala Lumpur, World's tallest building from 1998-2004; image © Jeff Goldberg/Esto

LYNN S. BEEDLE AWARD: PREVIOUS RECIPIENTS

2007

Lord Norman Foster

2006

Dr. Ken Yeang

2005

Dr. Alan Davenport

2004

Gerald Hines

2003

Charles DeBenedittis

2002

Dr. Lynn Beedle

CESAR PELLI

PELLI CLARKE PELLI ARCHITECTS

1984: Pelli's first high rise, Moma Residential Tower, New York City

1999: Cheung Kong Centre, Hong Kong; image © Tim Griffith/Esto

"An extraordinary career, Cesar Pelli continues to do great work. He has some of the world's tallest buildings to his credit – Petronas Towers and IFC to mention but two."

-Carol Willis , The Skyscraper Museum

AWARDS JURY / EDITOR STATEMENT

he is a Fellow of the American Institute of Architects, a Member of the American Academy of Arts and Letters, the National Academy of Design, the International Academy of Architecture, and of L'Academie d'Architecture de France. In addition, he has received over 200 awards for design excellence. In 1995, the American Institute of Architects awarded him the Gold Medal, its highest honor. In 2004, Mr. Pelli was awarded the Aga Khan Award for Architecture for the Petronas Towers.

2003: Two International Finance Center, Hong Kong; image © Tim Griffith/Esto

1991: One Canada Square (center) anchors Canary Wharf, London

CESAR PELLI

PELLI CLARKE PELLI ARCHITECTS

2005: Aerial view of Bloomberg Tower, New York City; image © Jeff Goldberg/ESTO

Awards Jury Statement:

Cesar Pelli has dramatically influenced the profession of architecture for decades. The relationship this has with the Lynn S. Beedle Award is obvious. Mr. Pelli's influence has been not only as an architect, but as an educator and statesman. This contribution has added significantly to the advancement of the profession of which we can all be grateful. Most notable are Mr. Pelli's designs which first and foremost engage the cities they are in and establish a new level of interest and excitement. The numerous designs for urban high rise buildings could be said to closely embody the aspirations of the CTBUH. Projects like the Petronas Towers in Kuala Lumpur, Malaysia—the tallest building in the world at the time of its completion—and the IFC in Hong Kong both draw from the inspiration of their place to create buildings that were innovative, unique, and advance the art and science of tall buildings.

AWARDS JURY / EDITOR STATEMENT

"Mr. Pelli's influence has been not only as an architect, but as an educator and statesman. This contribution has added significantly to the advancement of the profession of which we can all be grateful."

-Tim Johnson, NBBJ

2008: Rendering of Torre de Cristal, Madrid

2013 (expected): Rendering of Transbay Transit Center and Tower, San Francisco

WINNER:

2008 FAZLUR RAHMAN KHAN MEDAL

WILLIAM F. BAKER
SKIDMORE OWINGS & MERRILL LLP

William F. Baker is the Structural and Civil Engineering Partner for the Chicago and London offices of Skidmore, Owings & Merrill LLP (SOM). Since joining the firm in 1981, Bill's scope of engineering projects has extended from designing structural systems for supertall buildings, to smaller, specialized structures. He became an SOM partner in 1996.

Throughout his distinguished career, he has dedicated himself to structural innovation—most notably in the design of tall buildings within the urban landscape. In 1999, his next-generation, stayed-mast structural system for the proposed 7 South Dearborn, a 610-meter (2,000-foot) tall, 118-story tower built on only one acre (*see page 106*), incorporated structural engineering advances that demonstrated how modern wind engineering and materials can be used to achieve slender towers of unprecedented height. He would later continue such innovation by developing the 'buttressed core' structural system for the Burj Dubai (*see right*), a system which, in conjunction with sophisticated wind engineering, makes it possible to construct skyscrapers of extreme elevation. Set for completion in 2009, the Burj Dubai will be the tallest building in the world in all four categories recognized by the Council on Tall Buildings and Urban Habitat (CTBUH) including architectural height, the highest occupied floor, roof height and pinnacle height (*see page 116 for more on CTBUH Height Criteria*). Baker continues to push the boundaries of form and science with its neighbor, the twisting, 73-story Infinity Tower (*see page108*). It exemplifies Baker's belief that the exterior form of a high rise should be a direct expression of its structural framework; where, in the best instances, great structural and architectural collaboration results in situations where one cannot describe the structure without the architecture, and the architecture without its structure. Additionally, his contributions to the 66-story Nanjing Greenland Financial Center (*see page 109*) and the developing Lotte Super Tower (*see page 106*) secure his international reputation as a structural innovator.

Baker has left his mark on his hometown of Chicago by spearheading the structural design of the 63-story AT&T Corporate Center and the 92-story Trump International Hotel and Tower (*see page 107*). At 415 meters (1,361 feet) tall, Trump stands

Future tallest building in the world, Burj Dubai, nearing completion (scheduled for 2009)

FAZLUR RAHMAN KHAN MEDAL: PREVIOUS RECIPIENTS

2007

Dr. Farzad Naeim

2006

Srinivasa 'Hal' Iyengar

2005

Dr. Werner Sobek

2004

Leslie E. Robertson

WILLIAM F. BAKER

SKIDMORE OWINGS & MERRILL LLP

1999: Model of 7 South Dearborn proposal, Chicago

2012 (expected): Rendering of Lotte Super Tower, Seoul

"Bill's work has resulted in buildings that simultaneously mesh engineering and architecture to create the building form. This aspect of design innovation and collaboration was key to the jury. This was very much in the spirit of Khan and the great buildings that he was involved with at SOM."

-Dan Sesil, Leslie E. Robertson Associates

AWARDS JURY / EDITOR STATEMENT

as the second tallest building in North America, after the Sears Tower, and will be the second tallest concrete building in the world behind the Burj Dubai.

In addition to working at SOM, Baker is committed to research and education. His knowledgeable expertise is frequently solicited by institutions of higher learning, as well as numerous professional organizations. Most notably, he was engaged by the U.S. Federal Government in the wake of the 9/11 terrorist attacks to assist in understanding the World Trade Center collapse. As a scholar, he is currently focused on developing progressive research regarding the structural topology of tall buildings based on optimization methods. Baker is an adjunct thesis professor at the Illinois Institute of Technology College of Architecture, a fellow of the American Society of Civil Engineers (ASCE) and is on the Specifications Committee of the American Institute of Steel Construction (AISC). He frequently lectures on a variety of structural engineering topics within the U.S. and abroad.

2009 (expected): Trump International Hotel and Tower structurally topped out, viewed from Wabash Avenue

WILLIAM F. BAKER
SKIDMORE OWINGS & MERRILL LLP

2010 (expected): Rendering of Infinity Tower, Dubai

Awards Jury Statement:

The jury emphasized the significance of the Fazlur Khan award when it debated the nominees. The key aspects of this award were the "demonstrated excellence in design and research that has directly contributed to the advancement of tall building design." Bill Baker was selected as the award winner because of his extraordinary creativity in structural design and his integrated approach. This has resulted in buildings that simultaneously mesh engineering and architecture to create the building form. This aspect of design innovation and collaboration was key to the jury. This was very much in the spirit of Khan and the great buildings that he was involved with at SOM such as The John Hancock Tower and The Sears Tower. Mr. Baker's most recent contribution to the world of high rise design is the Burj Dubai, soon to be the tallest building in the world and a project that epitomizes the integration of structure and architecture. Through the early understanding of wind pressures, the building's simple and rational structural design significantly influenced the form of the building.

AWARDS JURY / EDITOR STATEMENT

"Bill Baker's structural design for the tallest building in the world, the Burj Dubai, is not only very simple and rational, but the building form itself is driven by the desire to minimize the wind pressures. Therefore the architect and engineer clearly worked hand in hand to create a very effective building. That to me is a key sustainable strategy."

-Tim Johnson, NBBJ

2009 (expected): Rendering of Nanjing Greenland Financial Center, Nanjing

Rendering of Yongsan Tower, a proposed 632-meter (2,073-foot) tower for Seoul

2008 CTBUH FELLOWS

SABAH AL-RAYES
PAN ARAB CONSULTING ENGINEERS

Sabah Mohammed Amin Al-Rayes founded Pan Arab Consulting Engineers (PACE) in 1968. He is a Senior Partner and Managing Director and is in charge of most of the tall buildings designed by PACE. He was Executive Secretary and member of the board of directors of the Kuwait Society of Engineers from 1967 to 1973.

He was the Chairman of the Education Committee for the Union of Arab Engineers in Cairo, Egypt from 1967 to 1984. He was a Member of the Board of Trustees for the College of Engineering at Kuwait University from 1977 to 1983. He is a member of the Global Advancement Council of Lehigh University and a fellow of the American Society of Civil Engineers.

He is the recipient of several honors for distinguished service, most notably during the Gulf War with the U.S. Army for the liberation of Kuwait. He has authored several publications including the Code of Ethics and professional Code of Practice for the Kuwait Society of Engineers. He also was the editor of the education newsletter of the Federation of Arab Engineers from 1980 to 1984 and editor of Traffic Safety Magazine from 1987 to 1995.

Sabah has been a long-standing friend and tireless promoter of the Council for more than three decades. A good friend of CTBUH founder Lynn Beedle, he has taken on many roles for the Council, including most recently serving as a member of the 2008 CTBUH 8th World Congress Organizing Committee, Dubai.

CTBUH STATEMENT

JAMES FORBES
HYDER CONSULTING PTY. LTD.

Jim Forbes is a Director of Hyder Consulting, an international advisory and design consultancy. Jim has been involved with planning, design, co-ordination and construction of a wide variety of buildings and engineering structures for over 40 years. His specific expertise in concrete structures led him towards a strong involvement in the Concrete Institute of Australia, and then the Federation International du Beton where he served as President from 2002 to 2004.

His project experience ranges from Tall Buildings both within Australia and around the world, to major public buildings such as Hospitals, Museums, Railway Stations and Theaters. Typical projects in which he has been involved include: the Australian Stock Exchange Building, Melbourne; Australia's New Parliament House, Canberra; Pikoi Tower, Hawaii; Southern Cross Station, Melbourne; Royal North Shore Hospital, Sydney; and proof engineer on Rialto Tower, Melbourne and Gateway Tower, Sydney. He has been involved with the CTBUH since 1990 and has served on the Steering Committee, the Height Committee and the Awards Committee, as well as being Structures Group Leader for many years. Jim was also Adjunct Professor of Civil Engineering at University of Sydney from 2005 to 2008.

Jim is a long-time supporter of the Council having served as co-chair of the original Building Systems and Concepts Group, overseeing eleven topical groups which produced several of the individual monographs published in the 1990's. He is currently working with the Brisbane-based Australian CTBUH to expand its reach to Sydney.

CTBUH STATEMENT

REVIEW OF THE 2007 AWARDS & AWARDS DINNER

Awards Committee Chairman, Tim Johnson of NBBJ, opens the ceremony

CTBUH Chairman David Scott (right) presents the Lynn S. Beedle Award to Foster + Partners Director, Michael Wurzel, on behalf of Lord Foster

For its sixth year, the CTBUH Annual Awards Dinner took a major step up in its scope and format in 2007—expanding in both the number of awards recognized and the number of people in attendance. Set within the atmospheric setting of Mies van der Rohe's iconic Crown Hall at the Illinois Institute of Technology, over 200 people joined us to celebrate the remarkable achievements of a number of distinguished individuals and project teams. Sold out several weeks before the event, table sponsors included Arup, Bovis Lend Lease, Goettsch & Partners, Halvorson & Partners, IIT College of Architecture, The John Buck Company, KONE, Perkins & Will, Schirmer Engineering, Solomon Cordwell Buenz, SOM, Thornton Tomasetti, Tishman Speyer, Walsh Construction and the WSP Group. The 30 people of the CTBUH international Steering Group were also in attendance, following the inaugural meeting of the group that day, and a visiting delegation of 15 colleagues from Russia completed the international flavor of the event.

Following cocktails and the formal dinner, Awards Committee Chairman Tim Johnson of NBBJ opened the presentation, welcomed guests, showcased the two gleaming steel sculptures his team at NBBJ had created for the awards and introduced the awards committee—Fiona Cousins of Arup, Jim Forbes of Hyder Consulting, James Goettsch of Goettsch & Partners, Vivian Loftness of Carnegie Mellon University and Dr. Shahzad Nasim of Meinhardt International.

CTBUH Chairman David Scott presented the first award—the Lynn S. Beedle Achievement Award to Lord Norman Foster of Foster + Partners. Unparalleled in terms of design excellence in tall buildings, Foster's track record spans from his seminal 1986 Hong Kong & Shanghai Bank Headquarters building Hong Kong, through the 1997 Commerzbank Frankfurt to more recent achievements such as the 2004 Swiss Re Tower London and the 2006 Hearst Tower New York. Michael Wurzel, a Director with Foster + Partners New York, collected the award on behalf of Lord Foster and shared a video message from Lord Foster to the audience. As noted by awards committee member Jim Forbes: "Lynn Beedle was a man who brought together art and science and truly facilitated integrated design. Lord Norman Foster clearly represents those traits as well".

The next award—the Fazlur Rahman Khan Medal—was presented by awards chair Tim Johnson to Dr. Farzad Naeim of John A. Martin & Associates. A registered Structural and Civil Engineer in the State of California, as well as an Attorney at Law and Member of the California Bar Association, Mr. Naeim has won numerous awards including the Los Angeles Tall Buildings Structural Design Council's Outstanding Journal Paper Award five times between 1996-2006. Citing the influence of Fazlur Khan on his development as a young engineer, Dr. Naeim shared with the audience how honored he was to receive the award. As noted by awards committee member Fiona Cousins: "Dr. Naeim has dedicated his life to a very focused topic pertinent to the high rise; seismic design. Although his focus has been narrow, its content and impact has tremendous depth".

The awards dinner venue, S.R. Crown Hall, Illinois Institute of Technology, Chicago (WSP / Beetham Organization / Ian Simpson Architects table in the foreground)

Best Tall Building Award—Beetham Hilton Tower, Manchester, UK

Beetham Tower team receiving Best Tall Building Award 2007 from Antony Wood, CTBUH Executive Director

Best Sustainable Tall Building Award —Hearst Tower, New York City

Brian Schwagerl of the Hearst Corporation with members of the project team receiving Best Sustainable Tall Building Award 2007

For the first time, in 2007, the two lifetime achievement awards were complemented with two new "building" awards which further the Council's ambition to seek to recognize both excellence in tall building design and the team achievement in successfully delivering any tall building project. The two awards—the "Best Tall Building Award" and "Best Sustainable Tall Building"—recognized a tall building that achieves sustainability at the highest and broadest level. (These would later become the foundation for the 2008 awards, in which the Council now recognizes four "Best Tall Building" awards by region and incorporates the sustainability criteria as a requirement for all.)

The recipient of the inaugural "Best Tall Building Award" went to the Beetham Hilton Tower, Manchester, UK—a stunning 169-meter (554-foot), 50-story building combining residential over hotel which, as the tallest residential building in the UK has, quite literally, changed the entire urban landscape of Manchester. As noted by awards chair Tim Johnson: "The Beetham Hilton Tower dramatically places itself in the historic context of Manchester. The building is sensitive to its urban surroundings and exudes an honest expression of its various programmatic functions. This is most notable in the powerful cantilever at the shift from hotel to residential". The award was presented by CTBUH Executive Director Antony Wood who spoke of his personal experiences of the building in his home city of Manchester. Mark Connolly of the Beetham Organization received the award, together with representatives of the project team—Ian Simpson architects, contractor Carillion and the WSP Group for structural, services and fire engineering.

The recipient of the inaugural "Best Sustainable Tall Building Award" went to the Hearst Tower, New York—a 182-meter (596-foot), 42-story office building for the Hearst Corporation which has unquestionably raised the bar for sustainable high rise design in North America. As noted by award committee member Vivian Loftness: "The Hearst Tower has many sustainable features, however, one that I am particularly fond of is that in discussions with building occupants they all express how they enjoy working in the building. Its air quality and day-lighting strategy are major contributors to this". The award was presented by Tim Johnson and received by Brian Schwagerl of the Hearst Corporation together with members of the project team—Foster & Partners architects, Turner Construction Managers, WSP Cantor Seinuk for structural and Flack & Kurtz for services.

Rounding off a thoroughly enjoyable evening was the recognition of two newly named CTBUH fellowships; Leslie E. Robertson and Chandra K. Jha.

CTBUH HEIGHT CRITERIA

The Council on Tall Buildings and Urban Habitat recognizes four different categories for the measuring of building height:

1. Height to Architectural Top

Height is measured from sidewalk level of the main entrance to the architectural top of the building, including spires, but not including antennae, signage or flag poles. This category is the most widely utilized and is used to define the rankings of the 100 Tallest Buildings in the World.

2. Height to Highest Occupied Floor

Height is measured from sidewalk level of the main entrance to the highest continually occupied floor within the building (i.e. not including maintenance areas).

3. Height to Top of Roof

Height is measured from sidewalk level of the main entrance to the highest point of the building's main roof level, not including spires or antennas.

4. Height to Tip

Height is measured from sidewalk level of the main entrance to the highest point of the building, irrespective of material or function of highest element (thus including antennae, flagpoles and signage).

World's Tallest Buildings According to **Height to Architectural Top**, as of Sept. 2008

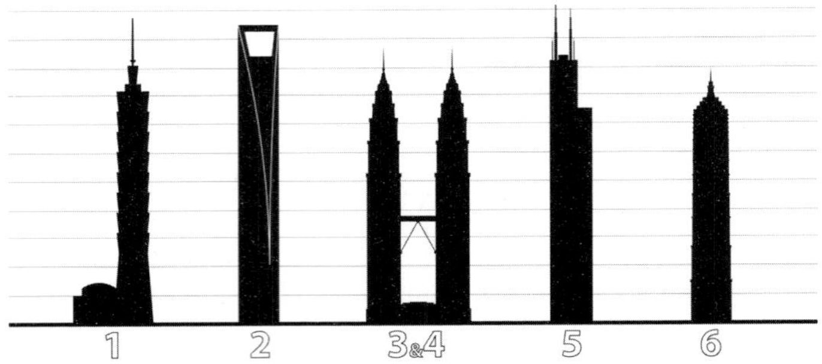

1. *Taipei 101, Taipei; 509 meters (1,670 feet)*
2. *Shanghai World Financial Center, Shanghai; 492 meters (1,614 feet)*
3-4. *Petronas Towers, Kuala Lumpur; 452 meters (1,483 feet)*
5. *Sears Tower, Chicago; 442 meters (1,451 feet)*
6. *Jin Mao Building, Shanghai; 421 meters (1,381 feet)*

World's Tallest Buildings According to **Highest Occupied Floor**, as of Sept. 2008

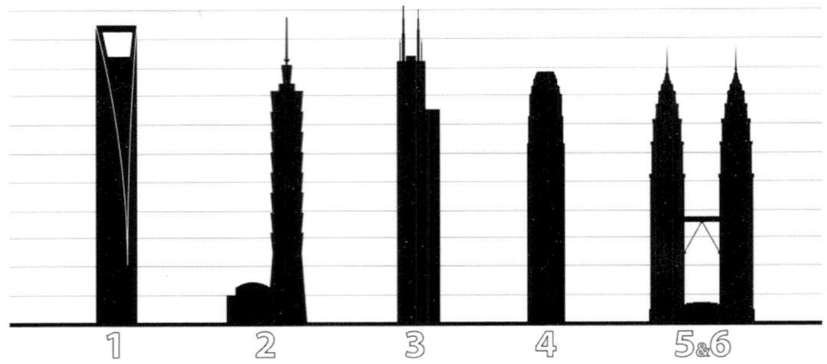

1. *Shanghai World Financial Center, Shanghai; 474 meters (1,555 feet)*
2. *Taipei 101, Taipei; 438 meters (1,440 feet)*
3. *Sears Tower, Chicago; 436 meters (1,431 feet)*
4. *Two International Finance Centre, Hong Kong; 377 meters (1,237 feet)*
5-6. *Petronas Towers, Kuala Lumpur; 375 meters (1,229 feet)*

World's Tallest Buildings According to **Height to Top of Roof**, as of Sept. 2008

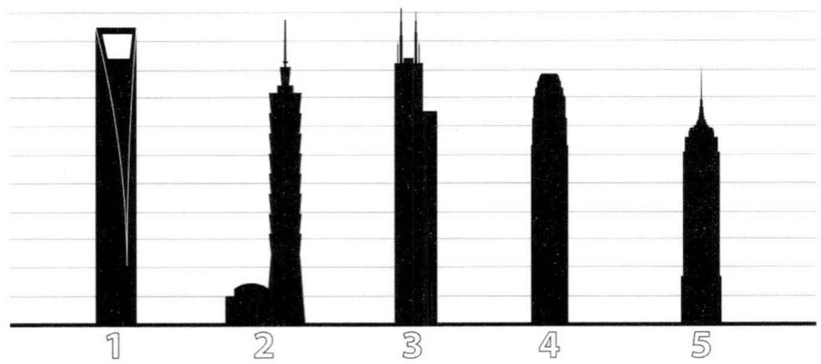

1. *Shanghai World Financial Center, Shanghai; 487 meters (1,599 feet)*
2. *Taipei 101, Taipei; 448 meters (1,473 feet)*
3. *Sears Tower, Chicago; 442 meters (1,451 feet)*
4. *Two International Finance Centre, Hong Kong; 392 meters (1,285 feet)*
5. *Empire State Building, New York City; 381 meters (1,250 feet)*

World's Tallest Buildings According to **Height to Tip**, as of Sept. 2008

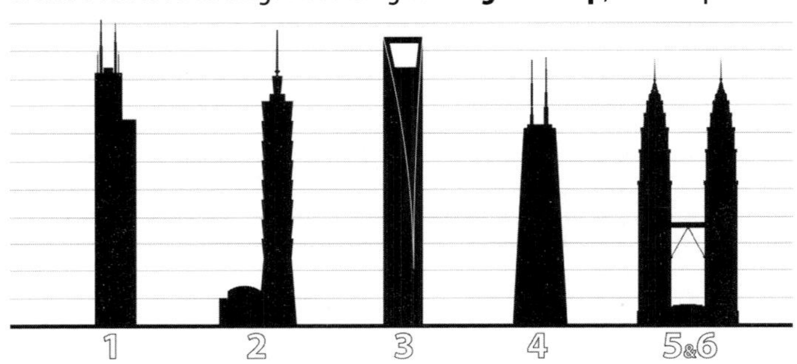

1. *Sears Tower, Chicago; 527 meters (1,730 feet)*
2. *Taipei 101, Taipei; 509 meters (1,670 feet)*
3. *Shanghai World Financial Center, Shanghai; 494 meters (1,622 feet)*
4. *John Hancock Center, Chicago; 456 meters (1,499 feet)*
5-6. *Petronas Towers, Kuala Lumpur; 452 meters (1,483 feet)*

100 TALLEST BUILDINGS IN THE WORLD

AS OF SEPT. 2008

The Council ranks the 100 tallest buildings in the world based on the height to architectural top (*see height criteria, left*). For a building to be given an official ranking it must be considered "complete".

The 100 Tallest Buildings list on the following pages also lists buildings that are currently under construction. (Although they will not be given a ranking until completed.) From reviewing this list the current boom in high rise construction quickly becomes apparent. By 2010 it is expected that 59 of the official "100 Tallest'" will be new compared with the list for 2006, just four short years earlier.

Color Key to Table Following:

Buildings listed in black are completed and officially ranked by the CTBUH.

Buildings listed in green are under construction and topped out.

Buildings listed in red are under construction, but not topped out.

Definitions Used for 100 Tallest Building List:

Construction Status:

A completed building can be considered such if it fulfils all three of the following criteria: a) topped out structurally and architecturally, b) fully-clad, and c) open for business, or at least partially occupied.

A tall building is considered "topped out" when all its structural members are in place.

A tall building is considered to be "under construction" when site clearing and excavation has been completed and foundations/piling work has begun.

Material:

A steel tall building is defined as one where the main vertical and lateral structural elements and floor systems are constructed from steel.

A concrete tall building is defined as one where the main vertical and lateral structural elements and floor systems are constructed from concrete.

A composite tall building utilizes a combination of both steel and concrete in the main structural elements throughout the building.

A concrete/steel tall building indicates a steel structural system located above a concrete structural system, with the opposite true of a steel/concrete building.

Use:

A single-function tall building is defined as one where 85% or more of its total floor area is dedicated to a single usage. A mixed-use tall building contains two or more functions, where each of the functions occupies at least 15% of the tower's total floor area. Ancillary/support areas such as car parks and mechanical plant space do not constitute mixed-use functions. Functions are denoted on CTBUH "Tallest" lists in ascending order.

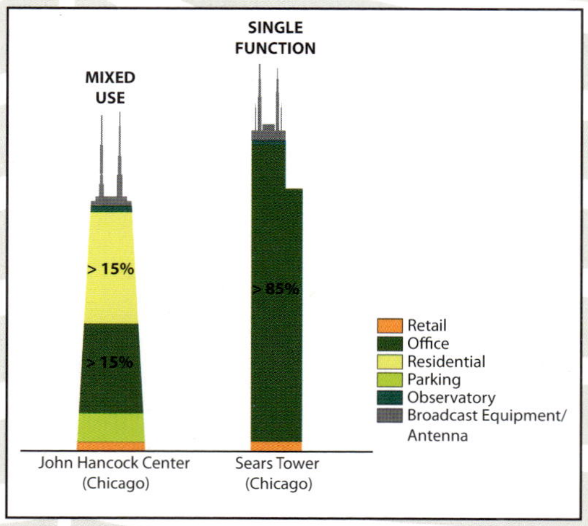

Diagram of single vs. mixed-use buildings

Rank	Building Name	City	Year	Stories	Height m	f	Material	Use
	Burj Dubai	Dubai	2009	160+	800+	2600+	Concrete/Steel	Hotel/Residential/Office
	Pentominium	Dubai	2012	120	618	2028		Residential
	Russia Tower	Moscow	2012	118	612	2009	Composite	Office/Hotel/Residential
	Chicago Spire	Chicago	2011	150	609	2000	Composite	Residential
	Mekkah Royal Clock Tower Hotel	Mekkah	2010	76	577	1893	Concrete	Hotel
	China 117 Tower	Tianjin	2013	117	570	1870		Office/Hotel/Residential
	Doha Convention Center Tower	Doha	2012	112	551	1808	Concrete/Steel	Hotel/Residential
	World Trade Center One	New York	2011	82	541	1776	Composite	Office
	Busan Lotte Tower	Busan	2013	107	510	1674	Composite	Office/Hotel
1	Taipei 101	Taipei	2004	101	509	1670	Composite	Office
	Burj Al Alam	Dubai	2011	108	501	1644	Composite	Office/Hotel
2	Shanghai World Financial Center	Shanghai	2008	101	492	1614	Composite	Office/Hotel
	International Commerce Centre	Hong Kong	2010	106	484	1588	Composite	Office/Hotel
3	Petronas Tower 1	Kuala Lumpur	1998	88	452	1483	Composite	Office
4	Petronas Tower 2	Kuala Lumpur	1998	88	452	1483	Composite	Office
	Nanjing Greenland Financial Center	Nanjing	2009	69	450	1476	Composite	Office/Hotel
5	Sears Tower	Chicago	1974	110	442	1451	Steel	Office
	Kingkey Finance Tower	Shenzhen	2011	97	439	1440		Office/Hotel
	Pearl River New City West Tower	Guangzhou	2009	104	438	1436	Composite	Office/Hotel
	Dubai Towers Doha	Doha	2010	84	437	1434	Concrete	Office/Hotel/Residential
6	Jin Mao Building	Shanghai	1999	88	421	1381	Composite	Office/Hotel
	Trump International Hotel & Tower	Chicago	2009	96	415	1362	Concrete	Hotel/Residential
7	Two International Finance Centre	Hong Kong	2003	88	415	1362	Composite	Office
	Princess Tower	Dubai	2009	101	414	1358	Concrete	Residential
	Marina 101	Dubai	2010	101	412	1352	Concrete	Hotel/Residential
	Al Hamra Tower	Kuwait City	2010	77	412	1352	Concrete	Office
	Emirates Park Towers Hotel & Spa 1	Dubai	2010	77	395	1296	Concrete	Hotel/Residential
	Emirates Park Towers Hotel & Spa 2	Dubai	2010	77	395	1296	Concrete	Hotel/Residential
8	CITIC Plaza	Guangzhou	1996	80	391	1283	Concrete	Office
	23 Marina	Dubai	2009	90	389	1276	Concrete	Residential
	Dalian Eton Center Tower 1	Dalian	2012	81	388	1274		
9	Shun Hing Square	Shenzhen	1996	69	384	1260	Composite	Office
	Central Market Residential Tower	Abu Dhabi	2010	88	382	1253		Residential
10	Empire State Building	New York	1931	102	381	1250	Steel	Office
	Elite Residence	Dubai	2010	87	380	1247	Concrete	Residential
	Mercury City Tower	Moscow	2009	70	380	1247	Composite	Office/Residential
	Sky Tower	Abu Dhabi	2009	83	379	1243	Concrete	Office/Residential
11	Central Plaza	Hong Kong	1992	78	374	1227	Concrete	Office
12	Bank of China	Hong Kong	1989	70	367	1205	Composite	Office
13	Bank of America Tower	New York	2008	54	366	1200	Steel	Office
	Vostok	Moscow	2010	95	365	1198	Concrete	Office
	Lam Tara Tower 1	Dubai	2010	70	360	1181	Concrete	Residential
14	Almas Tower	Dubai	2008	68	360	1181	Concrete	Office
15	Emirates Tower One	Dubai	1999	54	355	1165	Composite	Office
	The Square Capital Tower	Kuwait City	2010	63	351	1150		Office/Hotel
	D1 Tower	Dubai	2009	80	350	1148		Residential
	Gezhouba International Plaza	Wuhan		69	350	1148		Office
16	Tuntex Sky Tower	Kaohsiung	1997	85	348	1140	Composite	Office/Hotel

Rank	Building Name	City	Year	Stories	Height m	f	Material	Use
17	Aon Centre	Chicago	1973	83	346	1136	Steel	Office
	Faros de Panama Torre Central	*Panama City*	*2010*	*85*	*346*	*1135*	*Concrete*	*Office/Hotel/Residential*
18	The Center	Hong Kong	1998	73	346	1135	Steel	Office
	The Torch	*Dubai*	*2009*	*80*	*345*	*1132*	*Concrete*	*Residential*
19	John Hancock Center	Chicago	1969	100	344	1127	Steel	Office/Residential
	Torre Planetarium I	*Panama City*	*2011*	*92*	*343*	*1126*	*Concrete*	*Residential*
	Ahmed Abdul Rahim Al Attar Tower	*Dubai*	*2009*	*76*	*342*	*1122*	*Concrete*	*Residential*
	Megapolis Torre 2	*Panama City*	*2011*	*89*	*340*	*1116*	*Concrete*	
	The Wharf Times Square	*Wuxi*	*2011*	*83*	*339*	*1112*		*Multiple*
	Tianjin World Financial Center	*Tianjin*	*2010*	*76*	*337*	*1106*		*Office*
	Hanoi Gyoungnam Landmark Tower	*Hanoi*	*2011*	*70*	*336*	*1103*		*Office/Residential/Hotel*
	Parc1 Tower A	*Seoul*	*2011*	*72*	*334*	*1095*		*Office*
	Wenzhou Trade Center	*Wenzhou*	*2009*	*72*	*333*	*1094*	*Concrete*	*Office/Hotel*
20	Rose Rotana Tower	Dubai	2007	72	333	1093	Composite	Hotel
21	Shimao International Plaza	Shanghai	2006	60	333	1093	Concrete	Hotel
22	Minsheng Bank Building	Wuhan	2008	68	331	1087	Steel	Office
	**Ryugyong Hotel*	*Pyongyang*		*105*	*330*	*1083*	*Concrete*	*Hotel*
23	China World Trade Center Tower III	Beijing	2008	74	330	1083	Steel	Office/Hotel
	Hang Lung Plaza Tower 1	*Shenyang*	*2013*	*70*	*330*	*1083*		*Office/Hotel*
	The Index	*Dubai*	*2009*	*80*	*328*	*1076*	*Concrete*	*Office/Residential*
	Al Yaquob Tower	*Dubai*	*2009*	*69*	*328*	*1076*	*Concrete*	*Hotel/Residential*
	The Landmark	*Abu Dhabi*	*2009*	*72*	*324*	*1063*	*Concrete*	*Residential*
	Yantai Shimao No. 1 The Harbour	*Yantai*	*2010*	*57*	*323*	*1060*		*Office/Hotel*
24	Q1	Gold Coast	2005	78	323	1058	Concrete	Residential
25	Burj al Arab Hotel	Dubai	1999	60	321	1053	Composite	Hotel
	Hang Lung Plaza Tower 2	*Shenyang*	*2013*	*68*	*320*	*1050*		*Office/Hotel*
	Lam Tara Tower 2	*Dubai*	*2010*	*60*	*320*	*1050*	*Concrete*	*Office*
	Waterview Tower	*Chicago*	*2010*	*89*	*319*	*1047*	*Concrete*	*Hotel/Residential*
26	Nina Tower I	Hong Kong	2006	80	319	1046	Concrete	Office/Hotel
27	Chrysler Building	New York	1930	77	319	1046	Steel	Office
28	New York Times Tower	New York	2007	52	319	1046	Steel	Office
	HHHR Tower	*Dubai*	*2009*	*62*	*317*	*1040*	*Concrete*	*Residential*
29	Bank of America Plaza	Atlanta	1993	55	317	1039	Composite	Office
30	U.S. Bank Tower	Los Angeles	1990	73	310	1018	Steel	Office
	Ocean Heights	*Dubai*	*2009*	*82*	*310*	*1017*	*Concrete*	*Residential*
	Lamar Tower 1	*Jeddah*	*2010*	*65*	*310*	*1017*		*Office/Residential*
31	Menara Telekom Headquarters	Kuala Lumpur	1999	55	310	1017	Concrete	Office
	Pearl River Tower	*Guangzhou*	*2010*	*71*	*310*	*1016*	*Composite*	*Office*
32	Emirates Tower Two	Dubai	2000	56	309	1014	Concrete	Hotel
	City Hall and City Duma	*Moscow*	*2009*	*71*	*308*	*1012*		*Office*
	Infinity Tower	*Dubai*	*2010*	*73*	*307*	*1007*	*Concrete*	*Residential*
33	AT&T Corporate Center	Chicago	1989	60	307	1007	Composite	Office
	East Pacific Center Tower A	*Shenzhen*		*85*	*306*	*1004*		*Residential*
34	The Address Downtown Burj Dubai	Dubai	2008	63	306	1004	Concrete	Hotel/Residential
35	JP Morgan Chase Tower	Houston	1982	75	305	1002	Composite	Office

**Construction halted on the Ryugyong Hotel in 1992 and the project remained on hold until 2008, when construction on the tower resumed.*

Rank	Building Name	City	Year	Stories	Height m	f	Material	Use
	Torre Planetarium II	*Panama City*	*2011*	*82*	*305*	*1001*	*Concrete*	*Residential*
	NE Asia Trade Tower	*Incheon*	*2010*	*70*	*305*	*1001*	*Composite*	*Office/Hotel/Residential*
	Eurasia	*Moscow*	*2009*	*67*	*304*	*998*	*Composite*	*Office/Hotel/Residential*
36	Baiyoke Tower II	Bangkok	1997	85	304	997	Concrete	Hotel
37	Two Prudential Plaza	Chicago	1990	64	303	995	Concrete	Office
	Leatop Plaza	*Guangzhou*		*64*	*303*	*993*		*Office*
38	Wells Fargo Plaza	Houston	1983	71	302	992	Steel	Office
39	Kingdom Centre	Riyadh	2002	41	302	992	Concrete/Steel	Office/Hotel/Residential
	Gate of the Orient	*Suzhou*	*2009*	*68*	*302*	*990*		*Office/Hotel/Residential*
	India Tower	*Mumbai*	*2010*	*70*	*301*	*988*		*Hotel/Residential*
	Tianjin International Trade Tower	*Tianjin*	*2009*	*80*	*300*	*984*	*Steel*	*Office/Hotel*
	Gran Torre Costanera	*Santiago*	*2009*	*60*	*300*	*984*	*Concrete*	*Office*
	Arraya 2	*Kuwait City*	*2009*	*56*	*300*	*984*	*Concrete*	*Office*
40	Aspire Tower	Doha	2006	36	300	984	Composite	Office/Hotel
41	One Island East	Hong Kong	2008	69	298	979	Concrete	Office
42	First Bank Tower	Toronto	1975	72	298	978	Steel	Office
	Shanghai Wheelock Square	*Shanghai*	*2009*	*59*	*298*	*978*	*Concrete*	*Office*
43	Eureka Tower	Melbourne	2006	91	297	975	Concrete	Residential
44	Comcast Center	Philadelphia	2008	57	297	974	Composite	Office
45	Landmark Tower	Yokohama	1993	73	296	972	Steel	Office/Hotel
46	Emirates Crown	Dubai	2008	63	296	971	Concrete	Residential
	Khalid Al Attar Tower 2	*Dubai*	*2009*	*66*	*295*	*966*	*Concrete*	*Office/Residential*
	Central Park Office Tower	*Dubai*		*49*	*294*	*964*		*Office*
	Trump Ocean Club	*Panama City*	*2009*	*68*	*293*	*961*	*Concrete*	*Hotel/Residential*
47	311 South Wacker Drive	Chicago	1990	66	293	961	Concrete	Office
	Financial Tower	*Ho Chi Minh City*	*2009*	*61*	*292*	*958*	*Steel*	*Office*
48	SEG Plaza	Shenzhen	2000	71	292	957	Concrete	Office/Hotel
49	American International Building	New York	1932	67	290	952	Steel	Office
	TIPS China Building	*Shanghai*	*2009*	*67*	*290*	*951*		*Office*
	Chongqing Poly Tower	*Chongqing*		*60*	*290*	*951*		*Multiple*
	I&M Tower	*Dubai*	*2009*	*54*	*290*	*951*		*Office*
	Park Hyatt Guangzhou	*Guangzhou*	*2010*	*66*	*289*	*947*		*Office/Hotel/Residential*
50	Key Tower	Cleveland	1991	57	289	947	Composite	Office
51	Plaza 66	Shanghai	2001	66	288	945	Concrete/Steel	Office
	The Pinnacle	*London*	*2012*	*64*	*288*	*945*	*Steel*	*Office*
52	One Liberty Place	Philadelphia	1987	61	288	945	Steel	Office
	Excellence Century Plaza Tower 1	*Shenzhen*	*2010*	*60*	*288*	*945*		*Office*
	Capital City Moscow Tower	*Moscow*	*2009*	*73*	*286*	*938*	*Concrete*	*Office/Hotel/Residential*
	Moi City 1	*Shenyang*	*2010*	*70*	*286*	*938*		*Office*
	Sulafa Tower	*Dubai*	*2009*	*75*	*285*	*935*	*Concrete*	*Residential*
53	Millennium Tower	Dubai	2006	59	285	935	Concrete	Residential
54	Tomorrow Square	Shanghai	2003	55	285	934	Concrete	Office/Hotel/Residential
55	Columbia Center	Seattle	1984	76	284	933	Composite	Office
	G-Tower	*Dubai*	*2010*	*54*	*284*	*932*		*Residential*
56	Chongqing World Trade Center	Chongqing	2005	60	283	929	Concrete	Office
	Vision	*Brisbane*	*2010*	*72*	*283*	*928*		*Residential*
57	Cheung Kong Centre	Hong Kong	1999	63	283	928	Steel	Office

Rank	Building Name	City	Year	Stories	Height m	f	Material	Use
58	The Trump Building	New York	1930	71	283	927	Steel	Office
	Trump International Hotel & Tower	*Toronto*	*2010*	*57*	*282*	*925*	*Composite*	*Hotel/Residential*
	Catic Plaza	*Shenzhen*	*2009*	*58*	*281*	*922*		*Office*
	Torre Vitri	*Panama City*	*2010*	*75*	*281*	*921*	*Concrete*	*Residential*
59	Bank of America Plaza	Dallas	1985	72	281	921	Composite	Office
60	United Overseas Bank Plaza	Singapore	1992	66	280	919	Steel	Office
61	Republic Plaza	Singapore	1995	66	280	919	Composite	Office
62	Overseas Union Bank Centre	Singapore	1986	63	280	919	Steel	Office
	Greenland Square	*Zhengzhou*	*2009*	*63*	*280*	*919*		*Office/Hotel*
	Central Market Office Tower	*Abu Dhabi*	*2010*	*60*	*280*	*919*		*Office*
	Dalian Eton Center Tower 2	*Dalian*		*60*	*279*	*916*		
63	Citigroup Center	New York	1977	59	279	915	Steel	Office
64	Hong Kong New World Tower	Shanghai	2002	61	278	913	Composite	Retail/Office/Hotel
65	Diwang International Commerce Center	Nanning	2006	54	276	906	Concrete	Office/Hotel
66	Scotia Plaza	Toronto	1989	68	275	902	Composite	Office
	MAG 218 Tower	*Dubai*	*2009*	*66*	*275*	*902*	*Concrete*	*Residential*
	The Point	*Panama City*	*2009*	*65*	*275*	*902*	*Concrete*	*Residential*
67	Williams Tower	Houston	1983	64	275	901	Steel	Office
68	Wuhan World Trade Tower	Wuhan	1998	60	273	896		Office
	Lujing Tower	*Shenzhen*		*56*	*273*	*896*	*Concrete*	
	Parc1 Tower B	*Seoul*	*2011*	*59*	*271*	*889*		*Office*
	Faros de Panama Torre 2	*Panama City*	*2009*	*73*	*270*	*886*	*Concrete*	*Residential*
	Faros de Panama Torre 3	*Panama City*	*2009*	*73*	*270*	*886*	*Concrete*	*Residential*
69	The Cullinan I	Hong Kong	2008	68	270	886	Concrete	Residential
70	The Cullinan II	Hong Kong	2008	68	270	886	Concrete	Hotel/Residential
	Nikko Hotel Dubai	*Dubai*	*2009*	*60*	*270*	*886*	*Concrete*	*Hotel*
71	Renaissance Tower	Dallas	1975	56	270	886	Steel	Office
72	China International Center Tower B	Guangzhou	2007	62	269	884	Concrete	Office
73	Dapeng International Plaza	Guangzhou	2006	56	269	883	Composite	Office
74	21st Century Tower	Dubai	2003	55	269	883	Concrete	Residential
75	One Lujiazui	Shanghai	2008	47	269	883	Concrete	Office
76	Naberezhnaya Tower C	Moscow	2007	61	268	881	Composite	Office/Residential
	Lanko International Complex Yage Tower	*Chongqing*	*2008*	*54*	*268*	*879*	*Concrete*	*Office*
	Diamond of Istanbul	*Istanbul*	*2009*	*53*	*268*	*879*	*Composite*	*Office/Hotel/Residential*
	Beekman Tower	*New York*	*2010*	*76*	*267*	*876*	*Concrete*	*Residential*
	Charigali Tower	*Kuala Lumpur*	*2010*	*58*	*267*	*876*		*Office/Residential*
77	Al Faisaliah Center	Riyadh	2000	30	267	876	Concrete/Steel	Office
	The River Tower A	*Bangkok*	*2010*	*73*	*266*	*872*		*Residential*
78	900 North Michigan Avenue	Chicago	1989	66	265	871	Steel/Concrete	Retail/Hotel/Office/Residential
79	Bank of America Corporate Center	Charlotte	1992	60	265	871	Concrete	Office
80	Al Kazim Tower 1	Dubai	2008	53	265	869	Concrete	Residential
81	Al Kazim Tower 2	Dubai	2008	53	265	869	Concrete	Residential
82	BOCOM Financial Towers	Shanghai	1999	52	265	869		Office
83	120 Collins St.	Melbourne	1991	52	265	869	Concrete	Office
	WBC The Palace 1	*Busan*		*51*	*265*	*869*	*Concrete*	*Residential*
	WBC The Palace 2	*Busan*		*51*	*265*	*869*	*Concrete*	*Residential*
84	SunTrust Plaza	Atlanta	1992	60	264	867	Concrete	Office
	Arts Tower	*Panama City*	*2010*	*80*	*264*	*866*	*Concrete*	*Residential*

Rank	Building Name	City	Year	Stories	Height		Material	Use
					m	f		
85	Triumph Palace	Moscow	2005	61	264	866	Concrete	Residential
86	Shenzhen Special Zone Daily Tower	Shenzhen	1998	42	264	866	Concrete	Office
87	Tower Palace Three, Tower G	Seoul	2004	73	264	865	Composite	Residential
88	Trump World Tower	New York	2001	72	262	861	Concrete	Residential
89	Water Tower Place	Chicago	1976	74	262	859	Concrete	Retail/Hotel/Residential
90	Grand Gateway Plaza I	Shanghai	2005	54	262	859	Concrete	Office
91	Grand Gateway Plaza II	Shanghai	2005	54	262	859	Concrete	Office
92	Aon Center	Los Angeles	1974	62	262	858	Steel	Office
	Central Park Residential Tower	*Dubai*		*51*	*261*	*857*		*Residential*
	East Pacific Center Tower B	*Shenzhen*		*72*	*261*	*856*		*Residential*
93	Hotel Panorama	Hong Kong	2007	64	261	856	Concrete	Hotel/Residential
	Sapphire Tower	*Istanbul*	*2008*	*54*	*261*	*856*	*Concrete*	*Residential*
94	BCE Place-Canada Trust Tower	Toronto	1990	53	261	856	Composite	Office
	Marina Pinnacle	*Dubai*	*2009*	*67*	*260*	*853*	*Concrete*	*Residential*
	Hui Shang Center	*Changzhou*	*2010*	*63*	*260*	*853*		*Office*
	Vision Tower	*Dubai*	*2009*	*60*	*260*	*853*	*Concrete*	*Office*
	Shanghai International Finance Center I	*Shanghai*	*2011*	*60*	*260*	*853*	*Composite Office*	
	Albany Oasis Garden Office Tower	*Shanghai*	*2008*	*60*	*260*	*853*		*Office/Hotel*
95	Dual Towers 1	Manama	2007	57	260	853	Concrete	Office
96	Dual Towers 2	Manama	2007	57	260	853	Concrete	Office
97	101 Collins Street	Melbourne	1991	50	260	853	Composite	Office
98	Transamerica Pyramid	San Francisco	1972	48	260	853	Composite	Office
99	GE Building, Rockefeller Center	New York	1933	70	259	850	Steel	Office
100	Chase Tower	Chicago	1969	60	259	850	Steel	Office
100	Commerzbank Zentrale	Frankfurt	1997	56	259	850	Composite	Office

CTBUH ORGANIZATIONAL STRUCTURE

CTBUH Executive Committee
Chairman
David Scott
Executive Director
Antony Wood
Manager of Operations
Geri Kery
Immediate Past Chairman
Ron Klemencic

CTBUH Coordination
Systems
Charles Killebrew
Communications
Natalie Brush
Research
Philip Oldfield
CTBUH Journal Editor
Zak Kostura
Editorial Support
Robert Lau
Buildings Database Editor
Marshall Gerometta
Production
Katharina Holzapfel
Publications
Steven Henry
IT Support
Wai Sing Chew

CTBUH Steering Group
Ahmad K. Abdelrazaq, Samsung Corporation
Mir M. Ali, Univ. of Illinois, Urbana Champaign
Sabah Al-Rayes, Pan Arab Consulting Engineers
Carl Baldassarra, Schirmer Engineering Corporation
W. Gene Corley, CTL Group
Johannes de Jong, KONE International
Stephen V. DeSimone, DeSimone Consulting Engineers
Mahjoub Elnimeiri, Illinois Institute of Technology
James G. Forbes, Hyder Consulting
Thomas K. Fridstein, Hillier Architecture
Mayank Gandi, Remaking of Mumbai Federation
Louis F. Geschwindner, American Institute of Steel Construction
Timothy J. Johnson, NBBJ
Sang Dae Kim, Korea University
Jan Klerks, Dutch Council on Tall Buildings
Ryszard M. Kowalczyk, Univ. de Beira Interior
Simon Lay, WSP Group
Thomas J. McCool, Turner International LLC
Moira M. Moser, M. Moser Associates
Gary H. Pomerantz, Flack & Kurtz
Jerry R. Reich, Horvath Reich CDC, Inc.
Mark P. Sarkisian, Skidmore Owings and Merrill LLP
Simon Strzelecki, Bovis Lend Lease S.r.l.
Brett Taylor, Bornhorst + Ward Consulting Engineers
George von Klan, GVK-ECS, Inc.
Steve Watts, Davis Langdon LLP
Kenneth Yeang, Llewelyn Davies Yeang

CTBUH Working Groups Co-Chairs
Seismic Design
Ron Klemencic and Michael Willford
Progressive Collapse
Bob Smilowitz and Craig Gibbons
Fire & Safety
Simon Lay and Daniel O' Connor
Sustainable Design
Sadhu Johnson and Antony Wood
Construction Logistics / PM
David Crowell
Finance & Economics
Steve Watts and Dan Roberts

CTBUH Country Representatives
Argentina
Alberto Fainstein, AHFsa
Australia
Brett Taylor, Bornhorst + Ward Engineers
Austria
Ronald Mischek, Mischek Ziviltechniker GmbH
Canada
Barry Charnish, Halcrow Yolles
Finland (Latvia, Lithuania)
Matti-Esko Jarvenpaa, WSP Finland Ltd.
Greece
Alexios Vandoros, Vandoros & Partners
Hong Kong
Keith Griffiths, Aedas Ltd.
India
Lalit Gandhi, Remaking of Mumbai Federation
Korea
Sang-Dae Kim, Korea University
Malaysia
Faridah Shafii, Universiti Teknologi Malaysia
Netherlands
Jan Vambersky, Corsmit Raadgevend Ingenieursbureau BV
Philippines
Felino A. Palafox, Palafox Associates
Qatar
William Maibusch, Turner Construction International LLC
Russia
Tom McCool, Turner Construction International LLC
Saudi Arabia
Ibrahim Al Saudi, Saudi Oger Ltd.
Turkey
Husamettin Alper, Arup Mühendislik ve Müşavirlik Ltd. Şti.
United Arab Emirates
Andy Davids, Hyder Consulting Engineers
United Kingdom
Alastair Collins, Davis Langdon & Seah International
Vietnam
Nguyen Dinh Toan, The Architecture Research Institute

CTBUH ORGANIZATIONAL MEMBERS

Council on Tall Buildings and Urban Habitat

Supporting Contributors
Aedas Limited
Arup
Bovis Lend Lease S.r.l.
CB Richard Ellis Ltd
Davis Langdon LLP
DeSimone Consulting Engineers
Gale International
Illinois Institute of Technology, College of Architecture
Kohn Pedersen Fox
KONE Oyj
Mori Building Co., Ltd.
NBBJ
Remaking of Mumbai Federation
Samsung Corporation (Engineering & Construction)
Schindler Elevator Corporation
Schirmer Engineering Corporation
Turner
Webcor Builders
W.S. Atkins & Partners Overseas
WSP Group

Patrons
American Institute of Steel Construction, Inc.
BMT Fluid Mechanics Ltd.
DLA Piper UK LLP
Zuhair Fayez Partnership
Gifford Limited
Hongkong Land Ltd.
KLCC Property Holdings Berhad
Kuwait Foundation for the Advancement of Sciences
Saudi Ogers Ltd.
Skidmore Owings & Merrill, LLP
Tishman Speyer Properties

Donors
The Blume Foundation
Boundary Layer Wind Tunnel Laboratory (U. Western Ontario)
Built Form
Dagher Engineering
Halcrow Yolles
Halvorson and Partners
Hughes Associates, Inc.
Hyder Consulting
Rolf Jensen & Associates, Inc.
Larsen & Toubro Limited
MACE
Magnusson Klemencic Associates
W. L. Meinhardt Group Pty Ltd.
Walter P Moore
Nishkian Menninger Consulting and Structural Engineers
Omrania & Associates
Pan Arab Consulting Engineers
RISE International
Leslie E. Robertson Associates
RMJM Hillier
Rosenwasser/Grossman Consulting Engineers P.C.
Rowan Williams Davies & Irwin Inc.
Adrian Smith + Gordon Gill Architecture
Solomon Cordwell Buenz & Associates, Inc.
Studio Gang Architects
The Thornton Tomasetti Group
Weidlinger Associates, Inc.
Winward Structures Pty. Ltd.
Woods Bagot
Nabih Youssef & Associates

Contributors
AKF Engineers
American Iron and Steel Institute
Broadway Malyan
Canary Wharf Group
CB Engineers
Continental Automated Buildings Association (CABA)
Cook + Fox Architects LLP
CPP, Inc.
C.S. Structural Engineering, Inc.
DeStefano + Partners Ltd.
DHV Bouwen Industrie
Dong Yang Structural Engineers
Goettsch Partners
T. R. Hamzah & Yeang Sdn Bhd, Arkitek
INTEMAC
Japan Iron and Steel Federation
HP Konig, Heunisch und Partner
MulvannyG2 Architecture
Nakheel Tall Tower
NFPA
Norman Disney & Young
Pickard Chilton Architects, Inc.
Structal-Heavy Steel Construction
STS / AECOM
Studio Daniel Libeskind, Architect LLC

Participants
Al Jazera Consultants
Allford Hall Monaghan Morris LLP
APH Capital Partners
Arabian Aluminum Co. LLC
ARC Studio Architecture + Urbanism
Architectural Institute of Korea
Arquitectonica (Hong Kong)
Arquitectonica (US)
Axis Design Group Inc.
BG&E Pty Limited
Building Design International
Callison LLP
Canderel Management, Inc.
CBM Engineers
Chicago Committee on High-Rise Buildings
CS Associates
CTL Group
Dar Al-Handasah (Shair and Partners)
Dunbar & Boardman
Emirates Specialities Co. LLC
Terry Farrell & Partners
Fitzgerald Associates Architects
Flack + Kurtz
FXFOWLE Architects P.C.
Gold Coast City Council
GS Engineering & Construction
Gulf Glass Industries.
GVK Elevator Consulting Services, Inc.
Haynes Whaley Associates, Inc.
Heller Manus Architects
Hijjas Kasturi Associates Sdn.
Hilson Moran Partnership Ltd.
HOK,Inc.
Hong Kong Housing Authority
Horvath Reich CDC Inc.
International Union of Bricklayers and Allied Craftworkers
K.A.N. Development
Dennis Lau & Ng Chun Man, Architects & Engineers (HK) Ltd.
Lerch Bates Inc.
Limitless LLC
Stanley D. Lindsey & Associates
Lobby Agency
Lucien Lagrange Architects
Margolin Bros. Engineering & Consulting Ltd.
McNamara/Salvia, Inc.
Middlebrook & Louie
Miyamoto International, Inc.
Murphy/Jahn
Nikken Sekkei Ltd.
O'Connor Sutton Cronin
Palafox Associates
Pelli Clarke Pelli Architects
Plannungsgruppe Droge Baade Nagaraj
John Portman & Associates
Taylor Thomson Whitting Pty Ltd.
Tekla, Inc.
The Calvin Group, LLC
TPS Consult Ltd Centre
Traynor O'Toole Architects
University of Nottingham
Vanguard Realty Pvt. Ltd.
Vipac Engineers & Scientists Ltd.
Viracon
Walsh Construction Company
Werner Sobek Stuttgart GmbH & Co. KG
World Academy of Sciences for Complex Security
WSP Cantor Seinuk Group Inc.

**Supporting Contributors are those who contribute $10,000; Patrons: $6,000; Donors: $3,000; Contributors: $1,500; Participants: $750.*

INDEX BY BUILDING

181 William Street
Melbourne, Australia; 40

340 on the Park
Chicago, USA; 22

51 Lime Street
London, United Kingdom; 64

7 World Trade Center
New York City, USA; 24

AIG Tower
Hong Kong, China; 42

Aspire Tower
Doha, Qatar; 92

Bahrain World Trade Center
Manama, Bahrain; 86

Comcast Center
Philadelphia, USA; 26

Covent Garden
Brussels, Belgium; 70

Fujian Provincial Electric and Power Company Headquarters
Fuzhou, China; 44

Het Strijkijzer
The Hague, The Netherlands; 72

Kanyon Complex
Istanbul, Turkey; 74

Lumière Residences
Sydney, Australia; 46

Madou Plaza
Brussels, Belgium; 76

New York Times Building
New York City, USA; 16

Newton Suites
Singapore; 48

Nina Tower Development
Hong Kong, China; 50

One Island East
Hong Kong, China; 52

One Raffles Quay
Singapore; 54

One Rincon Hill
San Francisco, USA; 28

Park Tower at Beijing Yintai Centre
Beijing, China; 56

Pearl Breeze
Durban, South Africa; 94

Poly International Plaza
Guangzhou, China; 58

Sail @ Marina Bay
Singapore; 60

San Francisco Federal Building
San Francisco, USA; 30

Shanghai World Financial Center
Shanghai, China; 34

SyV Tower
Madrid, Spain; 78

Torre Espacio
Madrid, Spain; 80

Unity
Liverpool, United Kingdom; 82

INDEX BY COMPANY

AAARCHITECTEN b.v.; 73
Abssis; 71
AIG; 43, 61
Allford Hall Monaghan Morris LLP Architects; 83
AMEC Construction Management; 17
AM Jcd. Std.; 71
Aquilera Ingenieros; 79
AREP; 93
Archi 2000; 77
Architects 61 Pte Ltd.; 55
Arnold Associates; 53
Art & Build Architect sa; 71
Arup; 31, 41, 51, 53, 75, 93, 95
ASSAR Architects; 77
Atelier 50; 71
Atelier Ten; 27
Aveco De Bondt; 73
Axioma; 71

Bates Smart; 41
B.E. Greisch; 71
Beijing Urban Construction Group; 57
Beijing Yintai Property Co. Ltd.; 57
Belgometal; 71
Belt Collins; 61
BFBA Consultants; 95
Bijlbouw; 73
Boele & Van Esteren; 73
Bovis Lend Lease, Inc.; 23, 29
British Land Co Plc; 65
Buelens Real Estate; 71
Bureau Veritas; 79

CapitaLand; 43
Casa Design International Limited; 51
CB Engineers; 29
CBus Property; 41
Cerami and Associates, Inc.; 27
Charles Funke Associates; 65
Chinachem Group; 51
China State Construction Engineering Corporation; 35
Cicada Pte Ltd.; 49
City Development Ltd.; 61
Close; 71
Commission of the European Union; 77
Connell Mott MacDonald; 47
Cooper Architects; 95
Cornell Wagner; 61
Corsmit Consulting Engineers; 73
Cosentini Associates; 23
Creative Kingdom; 95

David Bonnett Associates; 65
Davis Langdon & Seah; 43, 61, 65
Deerns Consulting Engineers; 73
DEMOCO; 71
Dick Corporation; 31
Doha City; 93
Dragages Singapore - Bouygues Construction; 61

Ecart; 93
Eczacıbaşı Group; 75
Emmer Pfenninger; 65
Equation Lighting Design; 65
Estrumaher; 79

Faber Maunsell; 83
Forecast Investment Limited; 51
Forest City Ratner Companies; 17
Foster + Partners; 47, 65
Frasers Property; 47
Fujian Electric Power Co.; 45
FXFOWLE Architects; 17

Gensler; 17, 27
Geo Milieu; 71
Gordon Ingram Associates; 65
Goyne Adams; 83
Group Five; 95
Guangzhou Design Institute; 59
Guangzhou Gardens Planning & Building Design Institute; 59
Guangzhou No. 2 Construction & Engineering Co., Ltd.; 59

Hadi Simaan architects; 93
Heaton Associates; 27
Herpain Entreprise; 77
Hip Hing Construction Co. Ltd.; 43
Hoare Lea; 83
Hunt Construction Group; 31
Hyder Consulting Pty. Ltd.; 111

Immobiliere Du Royal Rogier; 71
Incorporated; 95
INTEMAC; 79
Irie Miyake Architects and Engineers; 35
İŞGYO of Turkey; 75
ITB; 71
IVG Real Estate Belgium; 77

James Carpenter Design Associates; 25
Jaros Baum and Bolles; 25
Jerde Partnership; 75
JNC International; 77
John Portman & Associates; 57
John Shreeves & Partners; 83
Joost Valgaeren; 71

Kajima Overseas Asia Pte Ltd.; 49
Keith Africa; 95
Kenchiku Setubi Sekkei Kenkyusho; 35
Kendall; 27
Ken Smith Landscape Design; 25
Ketlitz; 73
Kohn Pedersen Fox Associates PC; 35, 55
Korth Sunseri Hagey Architects; 29

Laing O'Rourke; 83
Lai Sun Development; 43
LBW Consultants; 49
Le Messurier; 57
Leslie E. Robertson Associates, R.L.L.P.; 35, 43
L.F. Discoll Co.; 27
Liberty Property Trust; 27
Lincolne Scott Ng Pte Ltd.; 49
LPA; 61

Mace Ltd.; 65
Magnusson Klemencic Associates; 23, 29
MC2 Estudio de Ingenieria; 79, 81

McKenzie Group Consulting; 41
Meinhardt Pte Ltd.; 53, 55, 61
Midmac-Six Construct JV; 93
Mitsubishi; 71
Montois Partners Architects sa; 71
Morganti General Contractors; 31
Mori Building Company Ltd.; 35
Morphosis; 31
Mott Green Wall; 65
Multiplex Constructions Pty Ltd.; 41, 47
MulvannyG2; 45
MVA (HK) Ltd.; 43

Nass, Murray & Roberts (JV); 87
NBBJ; 61
Newcomb & Boyd; 57
New York Times Company; 17
Norman Disney & Young; 41
Northeast China Architectural and Design Institute; 45

Obayashi Co.; 55
OHL Madrid; 81
Olin Partnership; 27
One Raffles Quay Pte Ltd.; 55
Oosthoek / Kemper; 73
Oostingh Staalbouw; 73

Pan Arab Consulting Engineers; 110
Pandev JV; 95
Parsons Brinckerhoff Ltd.; 57
Paul H. Yeomans, Inc.; 27
Pearls of Umhlanga Development (Pty) Ltd.; 95
Pei Cobb Freed & Partners Architects LLP; 81
Pelli Clarke Pelli Architects; 98
Pennoni Associates; 27
Philip Habib and Associates; 25
Placemedia Landscape Architects; 57
Poly Real Estate Group Co. Ltd.; 59
Precast Design Consultants; 61
PTW Architects; 47

Ramboll Whitbybird; 65
Reid Fenwick Associados; 81
Related Midwest; 23
Renzo Piano Building Workshop; 17
Rider Levett Bucknall Limited; 41, 53
Robert A.M. Stern Architects, LLP; 27
Robert Bird & Partners; 47
Roger Preston and Partners; 65
Rubio & Álvarez-Sala estudio de arquitectura; 79
Rumford Investments; 83
R. Urculo Ings Consultores Associados, S.A.; 81

SACYR; 79
Sandy Brown Associates; 65
Schlaich Bergermann und Partner; 79
Seco; 71
Seedat & Seedat Architects; 95
Shanghai Construction General Company; 35
Shiner + Associates, Inc.; 23
Sieben Energy Associates; 23
Silverstein Properties, Inc.; 25
SIP Project Managers; 95
SJB; 41
Skidmore, Owings & Merrill LLP; 25, 43, 59, 104
SmithGroup; 31
Sol Acoustics; 83
Solomon Cordwell Buenz; 23, 29
Stanhope Plc; 65
St Martins Property Group; 65
Strabag; 77
SWA Group; 59
Swire Properties Ltd.; 53

Tabanlıoğlu Architects; 75
Taylor Thomson Whitting; 47
Team Design; 61
Tepe Construction; 75
TESTA Inmuebles en Renta S.A.; 79
Thornton Tomasetti; 17, 27
Tishman Construction; 25
Torre Espacio Castellana, S.A.; 81
Tract Consultants; 47
Tractebel Development Engineering; 77
Turner Construction; 17
Twijnstra Gudde; 73
UOL Group Ltd.; 49
Urban West Associates; 29
Urbis JHD Pty Ltd; 41
Urculo Ingenieros; 79
U.S. General Services Administration; 31
Uys & White; 95

Van Den Rijse; 71
Van Deusen & Associates; 43
Van Roey; 71
Van Wetter; 71
Venac; 71
Vooruitzicht; 71

Walters & Simpson/LDM JV; 95
Waterman TCA; 77
Webcor Concrete; 29
Winward Structures; 41
WOHA Architects Pte Ltd.; 49
Wolff Associates; 23
Wong & Ouyang (HK) Ltd.; 53
WS Atkins & Partners; 87
WSP Cantor Seinuk; 25
WSP Flack + Kurtz; 17, 43, 59

IMAGE CREDITS

Cover: 51 Lime Street - © Nigel Young/Foster + Partners
The New York Times Building - © David Sundberg/Esto
Shanghai World Financial Center - © Mori Building Co., Ltd.
Bahrain World Trade Center - © WS Atkins & Partners

Pg 4-5: 51 Lime Street - © Nigel Young/Foster + Partners
The New York Times Building - © David Sundberg/Esto
Shanghai World Financial Center - © Mori Building Co., Ltd.
Bahrain World Trade Center - © WS Atkins & Partners
Cesar Pelli Portrait - © Pelli Clarke Pelli Architects
Bill Baker Portrait - © Skidmore, Owings & Merrill LLP/David Johnson Photo

Pg 12-13: 51 Lime Street image - © Nigel Young/Foster + Partners
The New York Times Building - © David Sundberg/Esto
Shanghai World Financial Center - © Mori Building Co., Ltd.
Bahrain World Trade Center image - © WS Atkins & Partners

Pg 14-15: 340 on the Park - © Solomon Cordwell Buenz
One Rincon Hill - © Magnusson Klemencic Associates
San Francisco Federal Building - © Tim Griffith
The New York Times Building - © David Sundberg/Esto
Comcast Center - © Peter Aaron/Esto
7 World Trade Center - © Vanni Archive/Ruggero Vanni

Pg 17: © David Sundberg/Esto

Pg 18-19: all exterior images © David Sundberg/Esto; all interior images © Nic Lehoux

Pg 20-21: all © FXFOWLE Architects

Pg 22-23: all © Solomon Cordwell Buenz

Pg 24-25: photo © Vanni Archive/Ruggero Vanni; drawings © Skidmore, Owings & Merrill LLP

Pg 26-27: all photos © Peter Aaron/Esto; all drawings © Robert A.M. Stern Architects, LLP

Pg 28-29: all © Magnusson Klemencic Associates

Pg 30: © Cody Andresen

Pg 31: all © Tim Griffith

Pg 32-33: One Raffles Quay - © Meinhardt Pte Ltd.
Shanghai World Financial Center - © Mori Building Co., Ltd.
Park Tower - © Beijing Yintai Centre
The Sail @ Marina Bay - © NBBJ
Newton Suites - © WOHA Architects Pte Ltd.
181 William - © Glenn Hester
Poly International Plaza - © Tim Griffith
Fujian Power Headquarters - © MulvannyG2
AIG Tower - © Skidmore, Owings & Merrill LLP
Lumière Residences - © Tom Evangelidis/Frasers Property
One Island East - © Wong & Ouyang (HK) Ltd.
Nina Tower Development - © Arup

Pg 35: © Kohn Pedersen Fox Associates PC

Pg 36: © Mori Building Co., Ltd.

Pg 37: photos © Kohn Pedersen Fox Associates PC; diagram © Leslie E. Robertson Associates, R.L.L.P.

Pg 38-39: photo © Mori Building Co., Ltd.; drawings & model © Kohn Pedersen Fox Associates PC

Pg 40-41: photo © Glenn Hester; drawing © Bates Smart; rendering © Flood Slicer P/L

Pg 42: © Skidmore, Owings & Merrill LLP

Pg 43: photo © AIG; rendering © Skidmore, Owings & Merrill LLP

Pg 44-45: all © MulvannyG2

Pg 46-47: photos © Tom Evangelidis/Frasers Property; drawing ©Foster + Partners/ PTW Architects/ Frasers Property

Pg 48-49: all © WOHA Architects Pte Ltd.

Pg 50: © Arup

Pg 51: all © Arup/Camera Vision Studio

Pg 52-53: all © Wong & Ouyang (HK) Ltd.

Pg 54-55: all © Meinhardt Pte Ltd.

Pg 56-57: photos/renderings © Beijing Yintai Centre; drawings © John Portman & Associates

Pg 58-59: all © Tim Griffith

Pg 60-61: all © NBBJ

Pg 62-63: 51 Lime Street - © Nigel Young/Foster + Partners
Unity - © Allford Hall Monaghan Morris LLP Architects/Tim Soar
SyV Tower - © Rafael Vargas
Torre Espacio - © Pei Cobb Freed and Partners
Het Strijkijzer - © Peter de Ruig
Madou Plaza - © ASSAR Architects
Kanyon Complex - © Arup
Covent Garden - © Art & Build Architects sa

Pg 65-68: all © Nigel Young/Foster + Partners

Pg 69: all © Foster + Partners

Pg 70-71: all © Art & Build Architects sa

Pg 72-73: exterior photos © Peter de Ruig ; interior photo © Luuk Kramer; drawing © **AA**Architecten bv

Pg 74-75: all © Arup

Pg 76-77: all © ASSAR Architects

Pg 78: © Rafael Vargas

Pg 79: photo © Roland Halbe; drawing © Rubio & Álvarez-Sala estudio de arquitectura

Pg 80-81: all © Pei Cobb Freed and Partners

Pg 82-83: all © Allford Hall Monaghan Morris LLP Architects/Tim Soar

Pg 84-85: Pearl Breeze - © Arup
Bahrain World Trade Center - © WS Atkins & Partners
Aspire Tower - © Midmac-Six Construct

Pg 87-91: all © WS Atkins & Partners

Pg 92-93: photos © Midmac-Six Construct ; rendering © Arup

Pg 94-95: all © Arup

Pg 98: © Pelli Clarke Pelli Architects

Pg 99: © Jeff Goldberg/Esto

Pg 100: Moma Residential Tower - © Cervin Robinson
Cheung Kong Center - © Tim Griffith/Esto

Pg 101: Two International Finance Centre - © Tim Griffith/Esto
One Canada Square - © Miller Hare Limited

Pg 102: © Jeff Goldberg/Esto

Pg 103: all © Pelli Clarke Pelli Architects

Pg 104: © Skidmore, Owings & Merrill LLP/David Johnson Photo

Pg 105: © Skidmore, Owings & Merrill LLP

Pg 106: 7 South Dearborn - © Skidmore, Owings & Merrill LLP/James Steinkamp Photography
Lotte Super Tower - © Skidmore, Owings & Merrill LLP/SWIM

Pg 107: © Skidmore, Owings & Merrill LLP/Lawrence C. Novak of the Portland Cement Association

Pg 108: © Skidmore, Owings & Merrill LLP/James Steinkamp Photography

Pg 109: Nanjing Greenland Financial Center - © Skidmore, Owings & Merrill LLP/James Steinkamp Photography
Yongsan Tower - © Skidmore, Owings & Merrill LLP/PMB Design